CAMBRIDGE LIBRARY COLLECTION

Books of enduring scholarly value

Darwin

Two hundred years after his birth and 150 years after the publication of 'On the Origin of Species', Charles Darwin and his theories are still the focus of worldwide attention. This series offers not only works by Darwin, but also the writings of his mentors in Cambridge and elsewhere, and a survey of the impassioned scientific, philosophical and theological debates sparked by his 'dangerous idea'.

Catalogue of the Osteological Portions of Specimens Contained in the Anatomical Museum of the University of Cambridge

William Clark was Professor of Anatomy at Cambridge for nearly 50 years, collecting many specimens of bones for use in the study of comparative anatomy, physiology and osteology. These formed a principal part of the collection that eventually became the university's Museum of Zoology. He wished to support students of natural sciences in acquiring knowledge from direct observation of well arranged and accurately identified specimens. The 1289 items, catalogued in 1862, include 128 from humans of varying races and dates. These include masks of the faces of Isaac Newton, William Pitt and Benjamin Franklin. This focus reflects, in part, the nineteenth-century fascination with phrenology. A regular participant in the influential Cambridge Philosophical Society, in May 1860 William Clark made there what Darwin perceived to be a "savage onslaught" on his recently published Origin of Species. This book reveals Clark's very different approach to studying the tree of life.

Cambridge University Press has long been a pioneer in the reissuing of out-of-print titles from its own backlist, producing digital reprints of books that are still sought after by scholars and students but could not be reprinted economically using traditional technology. The Cambridge Library Collection extends this activity to a wider range of books which are still of importance to researchers and professionals, either for the source material they contain, or as landmarks in the history of their academic discipline.

Drawing from the world-renowned collections in the Cambridge University Library, and guided by the advice of experts in each subject area, Cambridge University Press is using state-of-the-art scanning machines in its own Printing House to capture the content of each book selected for inclusion. The files are processed to give a consistently clear, crisp image, and the books finished to the high quality standard for which the Press is recognised around the world. The latest print-on-demand technology ensures that the books will remain available indefinitely, and that orders for single or multiple copies can quickly be supplied.

The Cambridge Library Collection will bring back to life books of enduring scholarly value across a wide range of disciplines in the humanities and social sciences and in science and technology.

Catalogue of the Osteological Portions of Specimens Contained in the Anatomical Museum of the University of Cambridge

WILLIAM CLARK

CAMBRIDGE
UNIVERSITY PRESS

CAMBRIDGE UNIVERSITY PRESS

Cambridge New York Melbourne Madrid Cape Town Singapore São Paolo Delhi

Published in the United States of America by Cambridge University Press, New York

www.cambridge.org
Information on this title: www.cambridge.org/9781108004671

This edition first published
This digitally printed version 2009

ISBN 978-1-108-00467-1

CATALOGUE

OF THE

ANATOMICAL MUSEUM,

CAMBRIDGE.

𝔈𝔞𝔪𝔟𝔯𝔦𝔡𝔤𝔢:
PRINTED BY C. J. CLAY, M.A.
AT THE UNIVERSITY PRESS.

𝔏𝔬𝔫𝔡𝔬𝔫: RIVINGTONS, 32, PATERNOSTER-ROW,
AND 3, WATERLOO PLACE.
𝔈𝔞𝔪𝔟𝔯𝔦𝔡𝔤𝔢: DEIGHTON, BELL, & CO.

CATALOGUE

OF THE

OSTEOLOGICAL PORTION OF SPECIMENS

CONTAINED IN

The Anatomical Museum

OF

THE UNIVERSITY OF CAMBRIDGE.

Printed for the Syndics of the University Press.

CAMBRIDGE:
AT THE UNIVERSITY PRESS.
1862.

PREFACE.

———•———

THIS Catalogue records the Osteological Specimens in the possession of the University. I have also included in it some Birds and Mammals stuffed, and some Fishes and Reptiles in spirit, because I think that it may frequently be advantageous to students, for whom this collection is chiefly designed, to have an opportunity of comparing the external characters of any groups in which they may be interested with the osteological. This is a plan which has been followed with success in several Continental Museums, especially in that of the University of Pavia.

As the value of such a collection as this depends in a great measure on the accuracy with which the specimens are named, I have never appended a specific name without due authority for it, and have preferred, in doubtful cases, to append the generic name alone. Whenever I found a difficulty in determining a specimen from descriptions, I took it to the British Museum, for comparison with a series of great extent and of different ages. I take this opportunity of thanking Professor Owen and Mr Gerrard for their great kindness in devoting many hours of their valuable time to the identification of the specimens shown to them.

A few remarks on the history of the Osteological Collection will not be inappropriate.

It originated in the private Museum of Sir Busick Harwood, Professor of Anatomy from 1785 to 1814, which the University purchased on his death. His specimens are not numerous, for his lectures were principally physiological. They are marked "Harwood collection."

In 1830 the University purchased a considerable part of the Museum of Joshua Brookes, Esq. Of this the osteological specimens are marked "Brookes collection."

In 1832 the collections were removed to the present buildings, and in 1836 the University purchased the whole of the valuable collection of Dr Macartney, Professor of Anatomy in Trinity College, Dublin. His specimens are marked "Macartney collection."

In 1853 I purchased in Paris of M. Dumoutier, who had accompanied the naturalists MM. Quoy and Gaimard on board the French frigate *Astrolabe*, which was absent in the Antarctic Seas on a scientific expedition from 1826—1829, a number of specimens, chiefly of the Seal tribe, collected by himself. These are marked [French Exp.].

In 1856 I had the pleasure of increasing the Collection by adding to it the osteological collection of Professor Bell, F.R.S., etc., etc., by which every order of Vertebrata is more adequately represented, and especially that of the Reptiles, amongst which is that valuable collection he had formed for the illustration of his work on the "Testudinata." His specimens are marked "Bell collection."

During the last two years the Collection has been entirely rearranged, the specimens already forming part of it cleaned, and in many cases rearticulated, numerous additions made, and the whole disposed in such a manner as to render it as accessible as our limited space will allow, for purposes of study. This it is hoped will be further facilitated by the publication of the Catalogue, which will also serve as a guide to persons desirous of adding to the collection by their own exertions or by purchase. It is worthy of remark, that as long as the very limited collection was deposited in the small dark room opposite Queens' College, where it remained till 1832, there was little inducement for either the Professor or any one else to add to it, as the specimens could not be exhibited. From the time of its removal to the present building until now it has steadily increased in the way above recorded, and by private donations. Now, however, a time has come when the old state of things is renewed, upon a larger scale it is true, and in a way less likely to attract notice, but still equally pernicious to the interests of Science. The space has become so inadequate that no further additions can be made, nor can several large skeletons, at present stowed in boxes, be exhibited at all. The University, for instance, possesses at the present time a Whale, purchased by subscription in 1850, an Eland, a Zebu Bull, a Red Deer, and a large species of Delphinus, not to mention a vast number of separate bones and smaller skeletons, which cannot be displayed even in fragments. Nor, supposing us willing to forego the display of these, and the acquisition of desiderata to fill up the *lacunæ* in the collection (for many orders, especially the Fish, are very inadequately represented), should we be blind to the fact that the Museum is now so crowded that it is difficult to have access to the skeletons; and that those persons who proceed to their degree in the Natural Sciences Tripos are thereby discouraged, and acquire the barely requisite amount of knowledge to enable them to satisfy the Examiners at secondhand from books, instead of from the actual specimens.

I cannot let this opportunity pass without tendering my thanks to Dr Drosier of Caius College, for his great kindness in assisting me in my lectures, when prevented by ill health from delivering them myself, and for the very beautiful skeletons of Birds, articulated by himself with great skill, with which he has enriched the Museum.

The Catalogue has been made, under my direction, by my son, who has also arranged the Collection, and articulated most of the recent additions.

It is hoped that the Catalogue of the Physiological Series of Specimens will shortly follow.

WILLIAM CLARK.

TRUMPINGTON STREET,
August, 1862.

CATALOGUE

OF

OSTEOLOGICAL SPECIMENS, ETC.

[N.B. The arrangement here adopted is that of Prof. VAN DER HOEVEN, as detailed in his *Handbook of Zoology:* and the numerals prefixed to the Orders and Families, when there are specimens to illustrate them, are in this Catalogue those prefixed to the same subdivisions in that work.]

CLASS I. FISHES. (*Pisces.*)

Order III. DESMIOBRANCHII.

Family IV. BATIDES.

.Genus *Myliobatis.* CUVIER.

Myliobatis. (*Species uncertain.*)

1. Teeth of one of the jaws. Presented by Prof. Clark.

Genus *Raia.* LINNÆUS.

Raia batis. L. **The Skate.**

2. Skeleton. Brookes collection.

Raia clavata. L.

3. Jaws of a female: to show the lozenge-shaped form of the teeth in a female. [Bell collection.] Presented by Prof. Clark.

Raia chagrinea. MONTAGU. **The Shagreen Ray.**

4. Cranium and teeth of a male. Harwood collection.

Genus *Pristis*. LATHAM.

Pristis antiquorum. LATH. **The Saw-fish.**

5. Beak of a very large specimen.
   Presented by Huddlestone Stokes, Esq.

6. Beak of a specimen nearly equal in size. [Bell collection.]
   Presented by Prof. Clark.

7. Beak of a small specimen. Macartney collection.

8. Beak of a smaller specimen. Macartney collection.

9. Beak of moderate size, wanting several teeth.
   Presented by Prof. Clark.

Family V. SELACHII.

Genus *Squatina*. DUMERIL.

Squatina vulgaris. RISSO. **The Angel Shark.**

10. Skeleton. Brookes collection.

11. Head, with the branchial arches. Presented by Prof. Clark.

12. The cartilaginous cranium. Presented by Prof. Clark.

Genus *Squalus*. LINNÆUS.

Squalus. (*Species uncertain.*) **Shark.**

13. Head. Harwood collection.

14. Jaws, with the teeth. [Bell collection.]
    Presented by Prof. Clark.

15. Sixty-seven vertebræ, articulated naturally.
    Presented by Mr Woolner, gardener of Downing College.

16. Thirty-eight vertebræ, articulated naturally.

17. Sixty-nine vertebræ, articulated naturally. [Bell collection.]
    Presented by Prof. Clark.

Genus *Selache*. CUVIER.

Selache. (*Species uncertain.*)

18. Part of the back-bone of a large species: some of the vertebræ
    are divided longitudinally to show the form of the articular sur-
    faces. Presented by F. Thackeray, M.D. Emmanuel College.

19. One of the vertebræ of the last specimen, divided transversely
    and macerated in order to show the radiated disposition of the
    osseous matter round the centre.

Genus *Carcharias*. CUVIER.

Carcharias. (*Species uncertain*.)

20.   Jaws, with the teeth.   Presented by Prof. Clark.

Order V. CHONDROSTEI.

Family VII. STURIONES.

Genus *Acipenser*. LINNÆUS.

Acipenser sturio. L. **The Sturgeon.**

21.   Head, without the pectoral fins.   [Bell collection.]
Presented by Prof. Clark.

22.   Head, with the osseous belt and the pectoral fins.   [Bell collection.]   Presented by Prof. Clark.

23.   The heterocercal tail, with its scutes, and the anal fin.   [Bell collection.]   Presented by Prof. Clark.

24.   A dermal scale of the back.   [Bell collection.]
Presented by Prof. Clark.

25.   The top of the head, formed by the expanded spines of the occipital and parietal vertebræ.   [Bell collection.]
Presented by Prof. Clark.

Order VII. LOPHOBRANCHII.

Family IX. LOPHOBRANCHII.

Genus *Pegasus*. LINNÆUS.

Pegasus draconis. L. **The Sea Dragon.**

26.   A dried specimen.   Presented by Prof. Clark.

Genus *Syngnathus*. LINNÆUS.

Syngnathus. (*Species uncertain*.) **Pipe-fish.**

27.   Five specimens preserved in spirit.   Presented by Prof. Clark.

28.   A dried specimen, apparently of the same species as the above.
Presented by Prof. Clark.

Genus *Hippocampus*. CUVIER.

Hippocampus brevirostris. CUV. **The Sea-horse.**

29. A specimen preserved in spirit.
                    Presented by W. Bayne, M.D. Trinity College.
30. Another specimen, similarly preserved.
31. Skeleton. Presented by Prof. Clark.
32. A dried specimen. Presented by Prof. Clark.
33. A dried specimen. Presented by Prof. Clark.

Order VIII. PECTOGNATHI.

Family X. GYMNODONTES.

Genus *Diodon*. LINNÆUS. **Urchin-fish.**
Diodon hystrix. L.

34. A stuffed specimen. Macartney collection.

Genus *Tetrodon*. LINNÆUS.

Tetrodon. (*Species uncertain.*)

35. A stuffed specimen. From the Indian Seas.
                    Presented by Huddlestone Stokes, Esq.

Family XI. SCLERODERMI.

Genus *Ostracion*. LINNÆUS.

Ostracion triqueter. L. **Coffer-fish.**

36. The dermal skeleton. Presented by Prof. Clark.
37. The same: from a young specimen. Presented by Prof. Clark.
38. The same: still smaller. Presented by Prof. Clark.

Order IX. MALACOPTERYGII.

Family XIII. CYPRINOIDEI.

Genus *Cyprinus*. LINNÆUS.

Cyprinus carpio. L. **The Carp.**

39. Skeleton. Brookes collection.
40. Skeleton. Presented by Prof. Clark.

Cyprinus brama. L. **The Bream.**

41. Skeleton. Presented by Prof. Clark.

Family XVIII. Esocii.

Genus *Esox.* LINNÆUS.

Esox lucius. L. **The Pike.**

42. Skeleton. Brookes collection.

Genus *Belone.* CUVIER.

Belone vulgaris. VALENC. **The Gar-fish or Sea-Pike.**

43. Head. Presented by Prof. Clark.

Genus *Exocœtus.* LINNÆUS. **Flying-fish.**

Exocœtus. (Species uncertain.)

44. A specimen preserved in spirit. Macartney collection.

Family XXVI. GADOIDEI.

Genus *Gadus.* LINNÆUS.

Gadus morrhua. L. **The Cod.**

45. Skeleton. Brookes collection.
46. Skeleton. Brookes collection.
47. Skeleton. Presented by Prof. Clark.
48. Head, articulated according to the vertebral system of Prof. Owen. [Bell collection.] Presented by Prof. Clark.
49. Head, similarly articulated. Presented by Prof. Clark.
50. The branchial arches, with the pharyngeal teeth. [Bell collection.] Presented by Prof. Clark.
51. Five caudal vertebræ. Presented by Prof. Clark.
52. Fifteen anterior vertebræ of the spinal column.
Presented by Prof. Clark.

Family XXVII. PLEURONECTÆ.

Genus *Hippoglossus.* CUVIER.

Hippoglossus vulgaris. FLEM. **The Holibut.**

53. Skeleton. Presented by Prof. Clark.
54. Head. Presented by Prof. Clark.

Genus *Platessa*. CUVIER.

Platessa vulgaris. FLEM. **The Plaice.**

55. Skeleton. Brookes collection.
56. Head. Presented by Prof. Clark.

Order X. ACANTHOPTERYGII.

Family XXXIII. HALIBATRACHI.

Genus *Lophius*. LINNÆUS.

Lophius piscatorius. L. **The Angler; Fishing Frog;**
or **Frog-fish.**

57. Skeleton. Brookes collection.
58. Bones of a skeleton complete, disarticulated.
Presented by Prof. Clark.
59. Head. Macartney collection.

Family XXXIV. BLENNIOIDEI.

Genus *Anarrhichas*. LINNÆUS.

Anarrhichas lupus. L. **The Wolf-fish. The Sea-Wolf.**

60. Head. Harwood collection.
61. Head, dried. [Bell collection.] Presented by Prof. Clark.

Family XXXV. GOBIODEI.

Genus *Callionymus*. LINNÆUS.

Callionymus lyra. L. **The Gemmeous Dragonet.**

62. Skeleton. From Mr Clayton, Fishmonger, Cambridge.
Presented by Prof. Clark.

Genus *Cyclopterus*. LINNÆUS.

Cyclopterus lumpus. L. **The Lump-fish.**

63. Skeleton. From Cambridge Market.
Presented by Prof. Clark.

Genus *Echeneis*. LINNÆUS.

Echeneis remora. L. **The Sucking-fish.**

64. Skeleton. Presented by Prof. Clark.

Family XXXVIII. SCOMBEROIDEI.

Genus *Lepidopus*. GOUAN.

Lepidopus argyreus. CUV. **The Scabbard-fish.**

65. Head, dried. Presented by Prof. Clark.

Genus *Xiphias*. LINNÆUS.

Xiphias gladius. L. **The Sword-fish.**

66. Head, with the pectoral fins. [Bell collection.]
Presented by Prof. Clark.

67. Head of a large specimen. [Bell collection.]
Presented by Prof. Clark.

68. Cranium and upper jaw. [Bell collection.]
Presented by Prof. Clark.

Genus *Zeus*. LINNÆUS.

Zeus faber. L. **The Dory.**

69. Skeleton. Macartney collection.
70. Head. Presented by Prof. Clark.

Family XLI. SCIÆNOIDEI.

Genus *Eques*. BLOCH.

Eques. (*Species uncertain.*)

71. Cranium, with two dorsal vertebræ: there is a remarkable swelling of the occiput. Macartney collection.

Family XLIII. ASPIDOPAREI.

Genus *Trigla*. LINNÆUS.

Trigla gurnardus. L. **The grey Gurnard.**

72. Skeleton. From Cowes, Isle of Wight.
Presented by Prof. Clark.

73. Skeleton. Brookes collection.

Family XLIV. PERCOIDEI.

Genus *Perca*. LINNÆUS.

Perca fluviatilis. L. **The Perch.**

74. Skeleton. Presented by Prof. Clark.

75.   Skeleton.
 Presented by H. J. H. Bond, M.D. Corpus Christi College.

76.   Portions of the bones of the head.   Do.

Perca labrax. L. The Basse, or Sea Perch.

77.   Skeleton.   Presented by Prof. Clark.

CLASS II.   REPTILES.   (*Reptilia.*)

Order II.   SAUROBATRACHI.

Family II.   PROTEOIDEA.

Genus *Hypochthon.*   MERREM.

Hypochthon Laurentii.   MERR.   *Proteus anguinus.*   AUCT.

78.   The animal preserved in spirit.
 Presented by Rev. Adam Sedgwick, M.A., Woodwardian Professor of Geology.

Genus *Menopoma.*   HARLAN.

Menopoma Alleghaniensis.   HARL.; formerly *Abranchus Alleg.* HARL.   It is called **Alligator,** or **Hell-bender** in America.

79.   Skeleton.   [Bell collection.]   Presented by Prof. Clark.

Order III.   BATRACHII.

Family IV.   BATRACHII.

Genus *Pipa.*   LAURENTI.

Pipa americana.   LAUR.

80.   The animal preserved in spirit.   Brookes collection.

Genus *Rana.*   LINNÆUS.

Rana esculenta. L. The Edible Frog.   The Green Frog.

81.   Skeleton from near Duxford, Cambridgeshire.
 Presented by G. E. Paget, M.D. Caius College.

Rana temporaria. L. **The Common Frog.**

82. Skeleton. Presented by Prof. Clark.
83. Skeleton of a large variety from Scotland. Taken in a marsh on the Pentland Hills. [Bell collection.]
Presented by Prof. Clark.

Genus *Hyla.* DUMERIL and BIBRON.

Hyla viridis. LAUR. **The Tree Frog.**

84. Several specimens in spirit. From Palermo, Sicily.
Presented by J. W. Clark, M.A. Trin. Coll.

Genus *Bufo.* LAURENTI.

Bufo vulgaris. LAUR. **The Common Toad.**

85. Skeleton. Presented by Prof. Clark.
86. Skeleton of a large variety from Jersey. Brookes collection.
87. Stuffed specimen of the same.
Presented by G. E. Paget, M.D. Caius College.

Bufo viridis. LAUR. **The Natter-Jack, or Natter-Jack Toad.**

88. Skeleton. Presented by Prof. Clark.

Order IV. OPHIDII.

Family V. VIPERINA.

Genus *Crotalus.* LINNÆUS.

Crotalus horridus. L. **The Rattle-Snake.**

89. Skeleton. Brookes collection.
90. Skeleton. Macartney collection.
91. Head. Harwood collection.

Genus *Vipera.* DAUDIN.

Vipera berus. DAUD. **The Viper. The Adder. The Aspic.**

92. Skeleton. Brookes collection.
93. Skeleton. Macartney collection.

2

Genus *Echidna*. MERREM.

Echidna arietans. MERR. **The Puff Adder.**

94. Skeleton. Brookes collection.

Family VI. ELAPINA.

Genus *Naia*. LAURENTI.

Naia tripudians. MERR. **The Cobra di Capello,** or **Hooded Snake.**

95. The animal in spirit.

Family IX. COLUBRINA.

Genus *Tropidonotus*. KUHL.

Tropidonotus torquatus. VAN DER HOEVEN.

Coluber natrix. L. **The Ringed Snake,** or **Common Snake.**

96. Skeleton. Brookes collection.
97. The animal in spirit.
    Presented by the Master and Fellows of Trinity College.
98. Head. Presented by Professor Clark.

Family XI. PYTHONINA.

Genus *Boa*. LINNÆUS.

Boa (Species uncertain.)

99. Head of a small specimen; there are no teeth in the pre-maxillary bone. Presented by Prof. Clark.

Genus *Python*. DAUDIN.

Python bivittatus. KUHL.

100. Skeleton. The animal died in Wombwell's menagerie.
    Presented by Prof. Clark.
101. Skin of the same, dried.

Python (Species uncertain), called *P. Apollinis* in Brookes' sale catalogue.

102. Skeleton. Brookes collection.

Family XVI. SCINCOIDEI.

Genus *Cyclodus*. WAGLER.

Cyclodus Boddaertii. DUM. and BIBRON.

Lacerta scincoides. WHITE. **The Common Scinc** of New Holland. DAUDIN.

103. Skeleton. [Bell collection.] Presented by Prof. Clark.
104. A stuffed specimen. Brookes collection.

Family XVIII. LACERTINI.

Genus *Varanus*. MERREM.

Varanus Niloticus. DUM. and BIBR. **The large Monitor.**

105. Skeleton. [Bell collection.] Presented by Prof. Clark.

Varanus Bengalensis. DUM. and BIBR.

106. The animal in spirit.
Presented by the Master and Fellows of Trinity College.

Family XIX. IGUANOIDEI.

Genus *Lophyrus*. DUMERIL.

Lophyrus tigrinus. DUM. and BIBR.

107. Stuffed specimen. Brookes collection.

Genus *Iguana*. DAUDIN.

Iguana tuberculata. LAUR.

108. Skeleton. Brookes collection.

Iguana nudicollis. CUV.

109. The animal in spirit.
Presented by the Master and Fellows of Trinity College.

Iguana (Species uncertain.)

110. Skeleton. [Bell collection.] Presented by Prof. Clark.
111. Skeleton. From the Zoological Society's Garden.
Presented by Prof. Clark.

Genus *Draco*. LINNÆUS.

Draco volans. L. **Flying Lizard,** or **Dragon.**

112. Skeleton. Presented by Prof. Clark.
113. Skeleton. [Bell collection.] Presented by Prof. Clark.
114. The animal in spirit. Presented by Prof. Clark.

Family XX. CHAMÆLEONIDEI.
Genus *Chamæleon.* AUCTORUM.

Chamæleon vulgaris. CUV.

115. Skeleton. Macartney collection.
116. Skeleton. [Bell collection.] Presented by Prof. Clark.

Family XXII. CROCODILINI.
Genus *Crocodilus.* SCHNEIDER.

Crocodilus sclerops. CUV. **Alligator,** or **Cayman.**

117. Skeleton. Macartney collection.
118. Young skeleton. Brookes collection.
119. Cranium and upper jaw of a large specimen. [Bell collection.]
Presented by Prof. Clark.
120. Young head. [Bell collection.] Presented by Prof. Clark.

Crocodilus vulgaris. CUV. **The Common Crocodile** of the Nile.

121. Young specimen in spirit. Brookes collection.
122. Head. From the Museum of J. P. Delafons, Esq.
Presented by Prof. Clark.
123. Head, articulated according to the vertebral system of Prof. Owen.

Crocodilus biporcatus. CUV.

124. Head. Presented by Mr Lichfield, Cambridge.
125. Head. [Bell collection.] Presented by Prof. Clark.
126. Young head. From Rive Matabanga, Kishnagur. [Bell collection.] Presented by Prof. Clark.

Crocodilus acutus. GEOFFROY ST HILAIRE.

127. Young skeleton. Brookes collection.

128. Young head. Brookes collection.

129. Head. The lower jaw has been injured during life, and an attempt made to repair it. [Bell collection.]
Presented by Prof. Clark.

Crocodilus Gangeticus. GMELIN. **The Gavial,** or more properly **"Gahrial."** **The Narrow-beaked Crocodile** of the Ganges.

130. Head, with the natural covering.
Presented to the Fitzwilliam Museum by the Rev. H. A. Bishop of St Catharine's College, and transferred thence in 1855 by order of the Vice-Chancellor.

131. Head. Brookes collection.

132. Young head. [Bell collection.] Presented by Prof. Clark.

133. Young head. [Bell collection.] Presented by Prof. Clark.

134. The nose. [Bell collection.] Presented by Prof. Clark.

135. Young specimen, stuffed.
Presented by Mr Haslop, Cambridge.

Crocodilus (Species uncertain.)

136. Head, wanting the intermaxillary bones, and the lower jaw. [French Exp.] From the Woody Islands, Borneo.
Presented by Prof. Clark.

Order VI. CHELONII.

Family XXIII. CHELONII.

Genus *Trionyx.* GEOFFROY ST HILAIRE. **Mud Tortoises.**

Trionyx labiatus. BELL.

137. Skeleton. [Bell collection.] Presented by Prof. Clark.

Trionyx Gangeticus. CUV.

138. Young skeleton. [Bell collection.] Presented by Prof. Clark.

Trionyx (Species uncertain.)

139. The carapace. Presented by Prof. Clark.

Genus *Emyda*. GRAY.

Emyda punctata. GRAY.

140.　Skeleton.　[Bell collection.]　Presented by Prof. Clark.

Genus *Emys*. BROGNIART. **The Terrapin.**

Emys tecta. BELL.

141.　The shell.　[Bell collection.]　Presented by Prof. Clark.

Emys picta. SCHWEIGGER.

142.　The animal in its shell, dried.　[Bell collection.]
Presented by Prof. Clark.

Emys rugosa. GRAY.

143.　Skeleton.　[Bell collection.]　Presented by Prof. Clark.

Genus *Hydraspis*. BELL.

Hydraspis (Species uncertain.)

144.　Skeleton.　[Bell collection.]　Presented by Prof. Clark.

Genus *Emysaura*. DUMERIL and BIBRON.

Emysaura serpentina. DUM. and BIBR.

145.　Skeleton.　(Bell collection.)　Presented by Prof. Clark.

Genus *Cinosternum*. SPIX.

Cinosternum scorpioides. GRAY. **The Pensylvanian Box Terrapin.**

146.　Skeleton.　[Bell collection.]　Presented by Prof. Clark.

Genus *Cistudo*. FLEMING.

Cistudo Carolinensis. GRAY. **The Box Terrapin.**

147.　Shell.　[Bell collection.]　Presented by Prof. Clark.
148.　Shell.　　　do.　　　　　　　do.
149.　Shell.　　　do.　　　　　　　do.

Cistudo Amboinensis. GRAY.

150.　Skeleton.　[Bell collection.]　Presented by Prof. Clark.

Genus *Testudo*. Brogniart. **Land Tortoises.**

Testudo græca. L.

151. Skeleton. Brookes collection.
152. Shell. [Bell collection.] Presented by Prof. Clark.
153. Shell. do. do.
154. Shell. do. do.
155. Shell, imperfect. Macartney collection.

Testudo indica. Gray.

156. Skeleton. Brookes collection.
157. Skeleton, with the sternum thrown back in order to display the internal arrangement of the bones. [Bell collection.]
Presented by Prof. Clark.

Testudo carbonaria. Spix.

158. Skeleton, imperfect. It shows the composition of the dorsal shield.
159. Skeleton, in its shell. The bones of the legs, head, and neck are covered with their natural integuments, dried. [Bell collection.] Presented by Prof. Clark.
160. Carapace, or dorsal shield. [Bell collection.]
Presented by Prof. Clark.
161. Plastron, or sternal shield. [Bell collection.]
Presented by Prof. Clark.

Testudo actinodes. Bell. **The Stellated Tortoise.**

162. Skeleton, imperfect in parts. The carapace appears to have been injured on the vertebral scutes, some of which have been removed. [Bell collection.] Presented by Prof. Clark.

Testudo pardalis. Bell.

163. Skeleton. [Bell collection.] Presented by Prof. Clark.

Testudo (Species uncertain.)

164. Head. [Bell collection.] Presented by Prof. Clark.
165. Head. do. do.

Genus *Chelonia.* B<small>ROGNIART.</small>

Chelonia imbricata. S<small>CHW.</small> **The Hawk's-bill Turtle.**

166. Skeleton. [Bell collection.] Presented by Prof. Clark.

167. Shell, with its scales, containing a portion of the skeleton. [Bell collection.] Presented by Prof. Clark.

Chelonia Mydas. S<small>CHW.</small> **The Green Turtle.**

168. Skeleton. [Bell collection.] Presented by Prof. Clark.

169. Head of a large specimen, with the skin turned back on each side to show the cavity of the tympanum. [Bell collection.]
Presented by Prof. Clark.

170. Head of a large specimen. [Bell collection.]
Presented by Prof. Clark.

171. Head. [Bell collection.] Presented by Prof. Clark.

172. Head.
173. Head, wanting the lower jaw. } [French Exp.] From Torres Strait, and the Gambier Islands. Presented by Prof. Clark.

174. Head, divided by a section in the mid-plane.
Presented by Prof. Clark.

175. Head, articulated according to the vertebral system of Prof. Owen. Presented by Prof. Clark.

Chelonia caouanna. S<small>CHW.</small> **The Logger-head Turtle.**

176. Head, wanting the lower jaw. [Bell collection.]
Presented by Prof. Clark.

177. Bones of the head, disarticulated. Presented by Prof. Clark.

Chelonia (Species uncertain.)

178. Head, dried. [Bell collection.] Presented by Prof. Clark.

179. The radius, ulna, carpus, metacarpal bones, and two phalangeal bones, of the left side: from a large specimen.
Presented by Prof. Clark.

180. The same, with three phalangeal bones, of the right side, from the same specimen. Presented by Prof. Clark.

Genus *Sphargis*. MERREM.

Sphargis coriacea. GRAY. **The Luth.**

Sphargis mercurialis. MERR.

181. Head. [Bell collection.] Presented by Prof. Clark.

182. Left anterior extremity of a large specimen. [Bell collection.]
Presented by Prof. Clark.

183. Fingers of the left anterior extremity of a smaller specimen.
[Bell collection.] Presented by Prof. Clark.

CLASS III. BIRDS. (*Aves*.)

Order 1. NATATORES.

Family I. BREVIPENNES.

Genus *Aptenodytes*. CUVIER.

Aptenodytes patagonica. FORSTER. **The Penguin.**

184. Skeleton. Brookes collection.

Genus *Alca*. LINNÆUS.

Alca torda. L. **The Razor-Bill.**

185. Head. Presented by J. W. Clark, M.A. Trin. Coll.
186. Head.   do.   do.
187. Head reversed, to show the inferior surface.
Presented by J. W. Clark, M.A. Trin. Coll.

Genus *Mormon*. ILLIGER.

Mormon fratercula. TEMM. **The Puffin.**

188. Skeleton. Macartney collection.
189. Head. Presented by J. W. Clark, M.A. Trin. Coll.
190. Head reversed, to show the inferior surface.
Presented by J. W. Clark, M.A. Trin. Coll.
191. Sternum, coracoid bones, clavicle, and scapulæ.
Presented by J. W. Clark, M.A. Trin. Coll.

Genus *Uria*. BRISSON.

Uria Troile. LATH. The Foolish Guillemot.

192. Skeleton. Macartney collection.
193. Head. Presented by J. W. Clark, M.A. Trin. Coll.
194. Head reversed, to show the inferior surface.
Presented by J. W. Clark, M.A. Trin. Coll.
195. Sternum, with the coracoid bones, and clavicle.
Presented by J. W. Clark, M.A. Trin. Coll.

Uria Brünnichii. SABINE. Brunnich's Guillemot.

196. Skeleton. The bird was taken by Mr Dunn, Naturalist, of Stromness in Orkney, on the North Coast of Iceland, in the summer of 1860. Presented by J. W. Clark, M.A. Trin. Coll.

Uria Grylle. LATH. The Black Guillemot.

197. Head. Presented by J. W. Clark, M.A. Trin. Coll.
198. Head reversed, to show the inferior surface.
Presented by J. W. Clark, M.A. Trin. Coll.
199. Sternum, clavicle, coracoid bones, and scapulæ.
Presented by J. W. Clark, M.A. Trin. Coll.

Uria lacrymans. TEMM. The Bridled Guillemot.

200. Head.
201. Sternum, coracoid bones, clavicle, and scapulæ.} From the Kalbaksfiord, Farö Islands.
Presented by J. W. Clark, M.A. Trin. Coll.

Genus *Colymbus*. LATHAM.

Colymbus septentrionalis. L. The Red-throated Diver.

202. Young skeleton.
Presented by W. H. Drosier, M.D. Caius College.

Colymbus arcticus. L. The Black-throated Diver.

203. Head.
204. Sternum, coracoid bones, clavicle, and scapulæ.}
Presented by W. H. Drosier, M.D. Caius College.

Colymbus glacialis. L. **The Great Northern Diver,** or **Ring-necked Loon.**

205. Skeleton of a bird of the second year.
Presented by W. H. Drosier, M.D. Caius College.

Colymbus (Species uncertain.)

206. Body, with the femora and tibiæ. Macartney collection.

Genus *Podiceps.* LATHAM.

Podiceps cristatus. LATH. **The Crested Grebe.**

207. Sternum, coracoid bones, clavicle, and scapulæ, of the female.
Presented by Mr Baker, Naturalist, Cambridge.

Podiceps minor. GMEL. **The Dabchick.**

208. Head. Presented by Prof. Clark.

Podiceps (Species uncertain.)

209. Sternum. Presented by Prof. Clark.

Family II. ANATINÆ.

Genus *Mergus.* LINNÆUS.

Mergus serrator. L. **The Red-breasted Goosander.**

210. Skeleton. Presented by Prof. Clark.
211. Sternum, coracoid bones, clavicle, and scapulæ, of the male, with the trachea from the mouth to its bifurcation.
Presented by Mr Baker, Naturalist, Cambridge.

Mergus Merganser. L. **The Buff-breasted Goosander.**

212. Sternum. Presented by J. W. Clark, M.A. Trin. Coll.

Genus *Oidemia.* FLEMING.

Oidemia nigra. FLEM. **The Black Scoter.**

213. Skeleton. Presented by W. H. Drosier, M.D. Caius College.
214. Head of an old male. Presented by Prof. Clark.

Oidemia fusca. FLEM. **The Velvet Scoter.**

215. Sternum, coracoid bones, clavicle, and scapulæ.
Presented by Mr Baker, Naturalist, Cambridge.

Genus *Somateria.* LEACH.

Somateria mollissima. LEACH. **The Eider Duck.**

216. Head. ⎫
217. Head reversed, to show the inferior surface. ⎬ From Stromness, Orkney.   Presented by J. W. Clark, M.A. Trin. Coll.

Genus *Clangula.* FLEMING.

Clangula chrysopthalma. STEPH. **The Golden-eyed Duck.**

218. Sternum, coracoid bones, clavicle, and scapulæ, of the male,
with the trachea, showing its two expansions.
Presented by Mr Baker, Naturalist, Cambridge.

219. Sternum, coracoid bones, clavicle, and scapulæ of a male.
Presented by Prof. Clark.

Genus *Fuligula.* STEPHENS.

Fuligula Marila. STEPH. **The Scaup Duck.**

220. Sternum, coracoid bones, clavicle, and scapulæ.
Presented by Mr Baker, Naturalist, Cambridge.

221. Head.   Presented by Prof. Clark.

Genus *Nyroca.* FLEMING.

Nyroca ferina. FLEM. **The Common Pochard.**

222. Head. ⎫
223. Head reversed, to show the inferior surface. ⎬
Presented by Prof. Clark.

Genus *Anas.* LINNÆUS.

Anas boschas. L. **The Wild Duck.**

224. Skeleton. [Bell collection.]   Presented by Prof. Clark.
225. Head. ⎫
226. Head reversed, to show the inferior surface. ⎬
Presented by Prof. Clark.

Anas, variety *domestica*. L.

227. Sternum.

228. Sternum, coracoid bones, clavicle, and scapulæ.}

Presented by Prof. Clark.

Anas arcuata. CUVIER.

229. A stuffed specimen. Presented by Thos. Horsfield, M.D.

Genus *Spatula*. BOIE.

Spatula clypeata. BOIE. **The Blue-winged Shovel-Bill.**

230. Young male, stuffed.

Presented by J. W. Clark, M.A. Trin. Coll.

231. Head. Presented by Prof. Clark.

Genus *Querquedula*. STEPHENS.

Querquedula crecca. STEPH. **The Teal.**

232. Skeleton. Presented by Prof. Clark.

233. Head.

234. Head reversed, to show the inferior surface.}

Presented by Prof. Clark.

235. Sternum, coracoid bones, clavicle, and scapulæ.

Presented by Mr Baker, Naturalist, Cambridge.

Genus *Tadorna*. LEACH.

Tadorna vulpanser. FLEM. **The Shiel Duck.**

236. Stuffed specimen of a male.

Presented by J. W. Clark, M.A. Trin. Coll.

237. Sternum, coracoid bones, clavicle, and scapulæ, of the male.

Presented by Mr Baker, Naturalist, Cambridge.

238. The same, of the female.

Presented by Mr Baker, Naturalist, Cambridge.

Genus *Mareca*. STEPHENS.

Mareca Penelope. SELBY. **The Widgeon.**

239. Head. Presented by Prof. Clark.

240. Sternum, coracoid bones, clavicle, and scapulæ.

> Presented by Prof. Clark.

241. Sternum. Presented by Prof. Clark.

Genus *Anser*. BRISSON.

Anser ferus. STEPH. **The Grey Goose.**

242. Head. Presented by J. W. Clark, M.A. Trin. Coll.

Anser, variety *domesticus.*

243. Head. Presented by Prof. Clark.

244. Head reversed, to show the inferior surface.

> Presented by Prof. Clark.

Genus *Bernicla*. STEPHENS.

Bernicla Brenta. STEPH. **The Brent Goose.**

245. Skeleton. Presented by W. H. Drosier, M.D. Caius College.

246. Head, from Kirkwall, Orkney.

> Presented by J. W. Clark, M.A. Trin. Coll.

Genus *Chenalopex*. STEPHENS.

Chenalopex Ægyptiacus. STEPH. **The Egyptian Goose.**

247. Sternum, coracoid bones, clavicle, and scapulæ. The bird was taken near Thirsk, Yorkshire.

> Presented by J. W. Clark, M.A. Trin. Coll.

Genus *Cygnus*. MEIJER.

Cygnus olor. GMEL. **The Swan.**

248. Skeleton. Presented by Prof. Clark.

249. Skeleton. Presented by Prof. Clark.

Cygnus ferus. RAY. **The Wild Swan,** or **Hooper.**

250. Skeleton. Presented by Prof. Clark.

251. Sternum, coracoid bones, clavicle, and scapulæ, with the whole of the trachea from the hyoid bone, which is attached to it, to its bifurcation.

252. Head.

253. Right foot, dried.

254. Left foot, dried.

The last four specimens are from an individual taken at the Laugavatn, Iceland. Presented by J. W. Clark, M.A. Trin. Coll.

255. Sternum, coracoid bones, clavicle, and scapulæ.
Presented by Prof. Clark.

256. The body, with a section in the keel of the sternum, to show the curvature of the trachea. Presented by Prof. Clark.

Cygnus (Species uncertain.)

257. Head. Presented by Prof. Clark.

Family III. STEGANOPODES.

Genus *Pelecanus.* ILLIGER.

Pelecanus Onocrotalus. BRUCH. **The Pelican.**

258. Skeleton. Presented by Prof. Clark.

259. Head, with the crop distended. [Bell collection.]
Presented by Prof. Clark.

260. Head, wanting the right tympanic and pterygoid bones.
Presented by Prof. Clark.

261. Head, wanting both the tympanic and pterygoid bones.
Presented by Prof. Clark.

Genus *Sula.* BRISSON.

Sula Bassana. BRISS. **The Solan Goose,** or **Gannet.**

262. Head. Presented by Prof. Clark.

Genus *Tachypetes.* VIEILLOT.

Tachypetes aquilus. VIEILL. **The Frigate-Bird.**

263. Head. Brookes collection.

Genus *Carbo*. Linnæus.

Carbo cristatus. Temm. **The Skart,** or **Green Cormorant.**

264.   Head.
265.   Head.
266.   Head reversed, to show the inferior surface.   } From Stromness, Orkney.

Presented by J. W. Clark, M.A. Trin. Coll.

Carbo Africanus. Lath.

267.   A stuffed specimen.   Presented by Thos. Horsfield, M.D.

Family IV. Longipennes.

Genus *Rhyncops*. Linnæus.

Rhyncops nigra. L.

268.   Head.   [Bell collection.]   Presented by Prof. Clark.

Genus *Sterna.* Linnæus.

Sterna Arctica. Temm. **The Arctic Tern.**

269.   Head.   From Loch Shell, Lewis.
           Presented by J. W. Clark, M.A. Trin. Coll.
270.   Head.   From Isle of Staffa.
           Presented by J. W. Clark, M.A. Trin. Coll.
271.   Sternum, coracoid bones, clavicle, and scapulæ. From the same specimen. Presented by J. W. Clark, M.A. Trin. Coll.

Genus *Larus.* Linnæus.

Larus Marinus. L. **The Greater Black-backed Gull.**

272.   Skeleton.   Presented by Prof. Clark.
273.   Head.   From the Isle of Skye.
           Presented by J. W. Clark, M.A. Trin. Coll.

Larus fuscus. L. **The Lesser Black-backed Gull.**

274.   Head.
275.   Head reversed, to show the inferior surface.
           Presented by J. W. Clark, M.A. Trin. Coll.

Larus tridactylus. LATH. **The Kittiwake.**

276. Head.
277. Head reversed, to show the inferior surface. }
                    Presented by J. W. Clark, M.A. Trin. Coll.

Larus argentatus. BRÜN. **The Herring-Gull.**

278. Head. From Basta Voe, Yell, Shetland.
                    Presented by J. W. Clark, M.A. Trin. Coll.

Larus minutus. PALL. **The Little Gull.**

279. The trunk. Presented by W. H. Drosier, M.D. Caius Coll.

Genus *Lestris.* ILLIGER.

Lestris catarractes. TEMM. **The Skua Gull.**

280. Head.
281. Head reversed, to show the inferior surface. } From Thors-havn, Farö Islands.
                    Presented by J. W. Clark, M.A. Trin. Coll.

Lestris Richardsonii. SWAINS. **Richardson's Skua.**

282. Head.
283. Head reversed, to show the inferior surface. }
                    Presented by J. W. Clark, M.A. Trin. Coll.

Genus *Puffinus.* RAY.

Puffinus Anglorum. RAY. **The Manx Shearwater.**

284. Head.
285. Head reversed, to show the inferior surface. }
                    Presented by J. W. Clark, M.A. Trin. Coll.

Genus *Procellaria.* LINNÆUS.

Procellaria glacialis. L. **The Northern Fulmar.**

286. Head, from the South of Iceland.
                    Presented by J. W. Clark, M.A. Trin. Coll.

4

26 BIRDS.

Procellaria (Species uncertain.)

287. Head, from the South Seas. [French Exp.]
Presented by Prof. Clark.

Genus *Thalassidroma.* VIGORS.

Thalassidroma pelagica. VIGORS. **The Storm Petrel.**

288. Skeleton. Presented by J. W. Clark, M.A. Trin. Coll.
289. Skeleton. do.

Genus *Diomedea.* LINNÆUS.

Diomedea exulans. L. **The Albatross.**

290. Head, dried, with the natural integuments.
Harwood collection.
291. Head. Harwood collection.
292. Skeleton. Brookes collection.
293. Head, from the Malvina or Falkland Islands. [French Exp.]
Presented by Prof. Clark.
294. Head, from the same locality. [French Exp.]
Presented by Prof. Clark.
295. Head, wanting part of the base of the skull.
Presented by Prof. Clark.

Diomedea melanophrys. TEMM.

296. Head, from the Falkland Islands. [French Exp.]
Presented by Prof. Clark.

Diomedea fuliginosa. GMEL. **The Sooty Albatross.**

297. Head. [French Exp.] Presented by Prof. Clark.

Order II. GRALLATORES.

Family V. MACRODACTYLI.

Genus *Fulica.* LINNÆUS.

Fulica atra. L. **The Bald Coot.**

298. Head. Presented by W. H. Drosier, M.D. Caius College.
299. Sternum, coracoid bones, clavicle, and scapulæ.
Presented by W. H. Drosier, M.D. Caius College.

Genus *Gallinula*. BRISSON.

Gallinula chloropus. LATH. **The Water-hen.**

300. Skeleton. Presented by Prof. Clark.
301. Head. do.

Gallinula phœnicurus. PENNANT.

302. A stuffed specimen. Presented by Thos. Horsfield, M.D.

Genus *Rallus*. LINNÆUS.

Rallus aquaticus. L. **The Water-Rail.**

303. Skeleton. Presented by Prof. Clark.
304. Head. do.
305. Head reversed, to show the inferior surface. Ditto.

Genus *Ortygometra*. LINNÆUS.

Ortygometra crex. L. **The Corn-crake.**

306. Head. Presented by Prof. Clark.

Family VI. LONGIROSTRES.

Genus *Scolopax*. LINNÆUS.

Scolopax rusticola. L. **The Woodcock.**

307. Head. Presented by Prof. Clark.
308. Head reversed, to show the inferior surface.
Presented by Prof. Clark.

Scolopax Gallinago. L. **The Common Snipe.**

309. Head. Presented by Prof. Clark.
310. Head reversed, to show the inferior surface.
Presented by Prof. Clark.

Scolopax Gallinula. L. **The Jack Snipe.**

311. Head. Presented by Prof. Clark.

Genus *Limosa*. BRISSON.

Limosa melanura. LEISL. The Black-tailed Godwit.

312.   Skeleton.   Presented by W. H. Drosier, M.D. Caius College.

Limosa rufa. BRISS. The Bar-tailed Godwit.

313.   Skeleton.   Presented by W. H. Drosier, M.D. Caius College.
314.   Bones of a skeleton, disarticulated.
                    Presented by W. H. Drosier, M.D. Caius College.
315.   Head.   Presented by Prof. Clark.

Genus *Totanus*. BECHSTEIN.

Totanus calidris. BECHST. The Redshank.

316.   Skeleton.   Presented by W. H. Drosier, M.D. Caius College.
317.   Head.   Presented by J. W. Clark, M.A. Trin. Coll.

Totanus affinis. HORSFIELD.

318.   A stuffed specimen.   Presented by Thos. Horsfield, M.D.

Genus *Tringa*. BRISSON.

Tringa minuta. LEISL. The Little Sandpiper.

319.   Skeleton.   Presented by W. H. Drosier, M.D. Caius College.

Tringa maritima. GMEL. The Purple Sandpiper.

320.   Head.
321.   Head reversed, to show the inferior surface. } From Iceland.
                    Presented by J. W. Clark, M.A. Trin. Coll.

Genus *Calidris*. ILLIGER.

Calidris arenaria. ILL. The Common Sanderling.

322.   Sternum, coracoid bones, clavicle, and scapulæ.
                    Presented by Mr Baker, Naturalist, Cambridge.

Genus *Phalaropus*. BRISSON.

Phalaropus lobatus. FLEM. The Grey Phalarope.

323.   Head, from Iceland.
                    Presented by J. W. Clark, M.A. Trin. Coll.

Genus *Numenius.* MOEHRING.

Numenius phæopus. LATH. **The Whimbrel.**

324. Skeleton. Presented by Prof. Clark.

325. Head. From Thorshavn. Farö Islands.
Presented by J. W. Clark, M.A. Trin. Coll.

326. Head reversed, to show the inferior surface. From the same locality. Presented by J. W. Clark, M.A. Trin. Coll.

327. A much smaller head.
Presented by J. W. Clark, M.A. Trin. Coll.

Family VII. CULTRIROSTRES.

Genus *Platalea.* LINNÆUS.

Platalea leucorodia. L. **The White Spoonbill.**

328. Skeleton. Presented by W. H. Drosier, M.D. Caius College.

329. Head. [Bell collection.] Presented by Prof. Clark.

Genus *Ciconia.* LINNÆUS.

Ciconia alba. BRISS. **The White Stork.**

330. Skeleton. Presented by Prof. Clark.

331. Skeleton of a male.
Presented by Mr Baker, Naturalist, Cambridge.

Ciconia (Species uncertain).

332. Left humerus. Harwood collection.

333. Right humerus. Harwood collection.

334. Sternum, coracoid bones, clavicle, and scapulæ. Harwood collection.

Ciconia marabou. TEMM. **The African Adjutant.**

335. Head. Presented by Prof. Clark.

Genus *Ardea.* LINNÆUS.

Ardea cinerea. L. **The Heron.**

336. Skeleton. Presented by J. W. Clark, M.A. Trin. Coll.

337. Head. Presented by Prof. Clark.

338. Head. do.

Ardea purpurea. L. **The Purple Heron.**

339. The legs. Presented by Thomas Horsfield, M.D.

Ardea speciosa. HORSFIELD.

340. A stuffed specimen. Presented by Thos. Horsfield, M.D.

Ardea cinnamomea. GMEL.

341. A stuffed specimen. Presented by Thos. Horsfield, M.D.

Genus NYCTICORAX. STEPHENS.

Nycticorax griseus. STEPH. **The Grey Night Heron.**

342. Sternum, coracoid bones, clavicle, and scapulæ. From a specimen shot near Malton, Yorkshire, May 26, 1857.
　　　　　　　Presented by J. W. Clark, M.A. Trin. Coll.

Genus *Grus.* LINNÆUS.

Grus Pavonina. PALL. **The Crested Crane.**

343. Skeleton. Brookes collection.
344. Head. Presented by Prof. Clark.

Genus *Phœnicopterus.* LINNÆUS.

Phœnicopterus antiquorum. TEMM. **The Flamingo.**

345. Skeleton, imperfect. From the Zoological Society's Gardens.
　　　　　　　Presented by Prof. Clark.
346. Head, dried. Presented by J. W. Clark, M.A. Trin. Coll.

Genus *Hæmatopus.* LINNÆUS.

Hæmatopus ostralegus. L. **The Oyster-catcher.**

347. Skeleton. Presented by W. H. Drosier, M.D. Caius College.
348. Head. From the Farö Islands.
　　　　　　　Presented by J. W. Clark, M.A. Trin. Coll.
349. Head reversed, to show the inferior surface.
　　　　　　　Presented by J. W. Clark, M.A. Trin. Coll.
350. A much smaller head. From the Isle of Skye.

Genus *Charadrius*. LINNÆUS.

Charadrius morinellus. L. **The Dotterel.**

351. Head. Presented by Prof. Clark.

352. Head reversed, to show the inferior surface.
Presented by Prof. Clark.

Charadrius pluvialis. L. **The Golden Plover.**

353. Head. Presented by Prof. Clark.

Genus *Squatarola*. CUVIER.

Squatarola cinerea. CUV. **The Gray Plover.**

354. Sternum, coracoid bones, clavicle, and scapulæ.
Presented by Mr Baker, Naturalist, Cambridge.

355. Another specimen. Presented by the same.

Genus *Œdicnemus*. TEMMINCK.

Œdicnemus crepitans. TEMM. **The Norfolk Plover.**

356. Skeleton. Presented by Prof. Clark.

Genus *Vanellus*. BRISSON.

Vanellus cristatus. MEYER. **The Crested Lapwing.**

357. Head. Presented by Prof. Clark.

358. Head reversed, to show the inferior surface.
Presented by Prof. Clark.

359. Head, with the eyes. Presented by Prof. Clark.

Vanellus tricolor. HORSFIELD.

360. A stuffed specimen. Presented by Thomas Horsfield, M.D.

Family IX. OTIDINÆ. **Bustards.**

361. Head, imperfect, of an unknown foreign species.
Presented by Prof. Clark.

Family X. PROCERI.

Genus *Dromaius*. VIEILLOT.

Dromaius novæ Hollandiæ. LATH. **The New Holland Ostrich, or Emeu.**

362. Skeleton, imperfect in the toes. Macartney collection.

363. Skeleton. [Bell collection.] Presented by Prof. Clark.

364. Left foot, dried. Harwood collection.

The following seven specimens were presented by the Cambridge Philosophical Society, in whose Museum the skin of the bird from which they were taken is preserved.

365. Right femur.

366. Left femur.

367. Pelvis, with the outer surface of the iliac bone on the left side removed to expose the vertebræ which compose the sacrum.

368. Sternum, with the coracoid bones and scapulæ.

369. The ten dorsal vertebræ.

370. The eighteen cervical vertebræ.

371. The ribs.

Genus *Casuarius*. LINNÆUS.

Casuarius galeatus. VIEILL. **The Asiatic Casuary.**

372. Left foot, dried. Harwood collection.

373. Left foot, dried.　　　　do.

Genus *Struthio*. LINNÆUS.

Struthio camelus. L. **The African Ostrich.**

374. Skeleton. From the Gardens of the City of London Zoological Society. Presented by Prof. Clark.

375. Head, articulated according to the vertebral system of Prof. Owen. Presented by Prof. Clark.

376. Left foot, dried. Harwood collection.

377. Left foot, dried.　　　　do.

378. Longitudinal section of the right femur, to show the air-cells. Harwood collection.

Genus *Dinornis*. OWEN.

Dinornis giganteus. OWEN.

The following four plaster casts were presented by the Royal College of Surgeons of England.

379. Left femur.

380. Left tibia, of the same individual.

381. Right tibia.

382. Left tarso-metatarsal bone.

Dinornis didiformis. OWEN.

383. Left tibia. Presented by Prof. Owen.

Dinornis casuarinus. OWEN.

384. Left femur. Presented by Prof. Owen.

385. Left tibia. do.

386. Left tarso-metatarsal bone. do.

Dinornis (Species uncertain).

387. Cast of the left femur.
Presented by the Royal College of Surgeons of England.

388. Casts of six phalangeal bones.
Presented by the Royal College of Surgeons of England.

Family XI. ALECTORIDES.

Genus *Palamedea*. LINNÆUS.

Palamedea cornuta. L. **The Kamichi,** or **American horned Screamer.**

389. The bones of the right wing, to show the spurs on the base and the head of the metacarpal bone. From a stuffed specimen.
Brookes collection.

390. The left wing, with its feathers, showing the osseous spines of the metacarpal bone. Brookes collection.

391. Left tibia, tarsus, and toes. Brookes collection.

Order III. GALLINÆ.

Family XIII. PENELOPINÆ.

Genus *Penelope.* MERREM.

Penelope marail. GMEL. **The Marail Guan.**

392. Skeleton. Brookes collection.

Family XIV. PHASIANINÆ.

Genus *Numida.* LINNÆUS.

Numida meleagris. L. **The Guinea Fowl.**

393. Head. Presented by Prof. Clark.
394. Head reversed, to show the inferior surface.
Presented by Prof. Clark.

Genus *Meleagris.* LINNÆUS.

Meleagris gallopavo. L. **The Turkey.**

395. Head. Presented by Prof. Clark.

Genus *Pavo.* LINNÆUS.

Pavo cristatus. L. **The Peacock.**

396. Skeleton. From the Zoological Society's Gardens.
Presented by Prof. Clark.

Genus *Phasianus.* LINNÆUS.

Phasianus colchicus. L. **The Pheasant.**

397. Skeleton. Presented by Prof. Clark.
398. Head. do.
399. Head reversed, to show the inferior surface.
Presented by Prof. Clark.

Genus *Gallus.* BRISSON.

Gallus Gallorum. LESS. **The Common Cock.**

400. Head. Presented by Professor Clark.
401. Head reversed, to show the inferior surface.
Presented by Professor Clark.

Gallus, var. *pentadactylus*. Temm. **The Dorking Fowl.**

402. Skeleton. Brookes collection.

Gallus, var. *cristatus*. Temm. **The Poland Fowl.**

403. Skeleton. Brookes collection.

Gallus furcatus. Temm.

404. A stuffed specimen. Presented by Thomas Horsfield, M.D.

Family XVI. Tetraoninæ.

Genus *Coturnix*. Moehring.

Coturnix vulgaris. Jardine. **The Quail.**

405. Head. Presented by Prof. Clark.

Genus *Perdix*. Brisson.

Perdix cinerea. Briss. **The Common Partridge.**

406. Head. Presented by Prof. Clark.
407. Head reversed, to show the inferior surface.
Presented by Prof. Clark.

Perdix rubra. Briss. **The Red-legged Partridge.**

408. Head. Presented by Prof. Clark.

Family XIX. Columbinæ.

Genus *Columba*. Linnæus.

Columba ænea. L.

408A. A stuffed specimen. Presented by Thomas Horsfield, M.D.

Columba palumbus. L. **The Wood Pigeon.**

409. Head. Presented by Prof. Clark.
410. Head reversed, to show the inferior surface.
Presented by Prof. Clark.

Columba risoria. L. **The Common Pigeon.**

411. Head. Presented by Prof. Clark.
412. Head reversed to show the inferior surface.
Presented by Prof. Clark.

Columba melanocephala. LATH.

413. A stuffed specimen. Presented by Thos. Horsfield, M.D.

Columba tigrina. TEMM.

414. A stuffed specimen. Presented by Thos. Horsfield, M.D.

Genus *Didus.* LINNÆUS.

Didus ineptus. L. **The Dodo.**

415. Cast in wax of the left foot of the Oxford specimen.
Presented by Mr Clark of Saffron Walden.

Order IV. SCANSORES.

Family XX. PSITTACINÆ.

Genus *Psittacus.* LINNÆUS.

Psittacus erythacus. L. **The Rose-billed Parakeet.**

416. Skeleton. Presented by Prof. Clark.
417. Skeleton. Macartney collection.
418. Head. Presented by Prof. Clark.

Psittacus domicella. L. **The Purple-capped Lory.**

419. Skeleton. Presented by Prof. Clark.

Psittacus Ponticereanus. LATH.

420. A stuffed specimen. Presented by Thos. Horsfield, M.D.

Psittacus (Species uncertain).

421. Skeleton. [Bell collection.] Presented by Prof. Clark.
422. Skeleton. Presented by Prof. Clark.

Genus *Macrocercus.* VIEILLOT.

Macrocercus Macao. VIEILL. **The Blue and Yellow Macaw.**

423. Skeleton. From the Zoological Society's Gardens.
Presented by Prof. Clark.

424. Head. From the Zoological Society's Gardens.
Presented by Prof. Clark.

425. Head reversed, to show the inferior surface.
Presented by Earl Fitzwilliam.

Macrocercus (Species uncertain).

426. Skeleton. [Bell collection.] Presented by Prof. Clark.

Genus *Centropus.* ILLIGER.
Centropus affinis. HORSFIELD.

427. A stuffed specimen. Presented by Thos. Horsfield, M.D.

Family XXII. POGONOPHORÆ.
Genus *Bucco.* LINNÆUS.
Bucco Javensis. HORSFIELD.

428. A stuffed specimen. Presented by Thos. Horsfield, M.D.

FAMILY XXIV. CUCULINÆ.
Genus *Cuculus.* LINNÆUS.
Cuculus canorus. L. **The Cuckoo.**

429. Skeleton. Presented by Prof. Clark.

FAMILY XXV. SAGITTILINGUES.
Genus *Picus.* LINNÆUS.
Picus viridis. L. **The Green Woodpecker.**

430. Skeleton. Presented by Prof. Clark.
Picus tiga. HORSFIELD.
431. A stuffed specimen. Presented by Thos. Horsfield, M.D.

Order V. PASSERINI.

Family XXVIII. BUCEROTINÆ.

Genus *Buceros*. LINNÆUS.

Buceros hydrocorax. GMEL.

432.   Head, dried. From Sumatra. [French Exp.]

Presented by Prof. Clark.

Family XXXI. HALCYONINÆ.

Genus *Alcedo*. LINNÆUS.

Alcedo ispida. L. **The Kingfisher.**

433.   Head. Presented by Prof. Clark.

434.   Head reversed, to show the inferior surface.

Presented by Prof. Clark.

The six following stuffed specimens were presented by Thomas Horsfield, M.D.:

435.   *Alcedo Meninting.* HORSFIELD.

436.   *Alcedo tridactyla.* LINNÆUS.

437.   *Alcedo melanoptera.* HORSFIELD.

438.   *Alcedo omnicolor.* TEMMINCK.

439.   *Alcedo colaris.* LATH.

440.   *Alcedo leucocephala.* GMELIN.

Family XXXIII. TROCHILIDÆ.

Trochilus (Species uncertain). **Humming-bird.**

441.   Head. [Bell collection.] Presented by Prof. Clark.

Family XXXVII. CORVINÆ.

Genus *Corvus*. LINNÆUS.

Corvus corax. L. **The Raven.**

442.   Skeleton. Presented by Prof. Clark.

443.   Head. From the Geysers, Iceland.

Presented by J. W. Clark, M.A. Trin. Coll.

444. Head reversed, to show the inferior surface. From the same locality. Presented by J. W. Clark, M.A. Trin. Coll.

445. Head. Presented by H. J. H. Bond, M.D. Corpus Christi Coll.

Corvus corone. L. **The Carrion Crow.**

446. Head. Presented by Prof. Clark.

Corvus cornix. L. **The Hooded, or Royston Crow.**

447. Head. Presented by J. W. Clark, M.A. Trin. Coll.

Corvus frugilegus. L. **The Rook.**

448. Head. Presented by Prof. Clark.

449. Head reversed, to show the inferior surface.
Presented by Prof. Clark.

450. Sternum, coracoid bones, clavicle, and scapulæ.
Presented by Prof. Clark.

Corvus monedula. L. **The Jackdaw.**

451. Skeleton. Presented by Prof. Clark.

452. Head. do.

453. Head. do.

Corvus albicollis. LATH. **The African Crow.**

454. Head. Brookes collection.

Genus *Crypsirhina.* VIEILLOT.

Crypsirhina Temmia. VIEILL.

455. A stuffed specimen. Presented by Thos. Horsfield, M.D.

Family XXXVIII. PARADISEINÆ.

Genus *Oriolus.* LINNÆUS.

Oriolus Cochinchinensis. BRISS.

456. Stuffed specimen. Presented by Thos. Horsfield, M.D.

457. Another specimen. do.

Family XXXIX. STURNINÆ.

Genus *Pastor*. TEMMINCK.

Pastor cristatellus. TEMM.

458. Stuffed specimen. Presented by Thos. Horsfield, M.D.

Genus *Sturnus*. LINNÆUS.

Sturnus vulgaris. L. **The Starling.**

459. Head. Presented by Prof. Clark.
460. Head reversed, to show the inferior surface.
Presented by Prof. Clark.

Family XL. FRINGILLINÆ.

Genus *Pyrrhula*. BRISSON.

Pyrrhula vulgaris. BRISS. **The Bullfinch.**

461. Skeleton. Presented by Prof. Clark.
462. Head. do.
463. Head reversed, to show the inferior surface.
Presented by Prof. Clark.

Genus *Fringilla*. LINNÆUS.

Fringilla domestica. L. **The Sparrow.**

464. Skeleton. Presented by Prof. Clark.
465. Head. do.
466. Head reversed, to show the inferior surface.
Presented by Prof. Clark.

Fringilla cœlebs. L. **The Chaffinch.**

467. Head. Presented by Prof. Clark.
468. Head reversed, to show the inferior surface.
Presented by Prof. Clark.

Fringilla chloris. TEMM. **The Greenfinch.**

469. Skeleton. Presented by Prof. Clark.

Fringilla Canarina. L. **The Canary.**

470. Skeleton. Presented by Prof. Clark.
471. Skeleton.   do.
472. Head.   do.
473. Head reversed, to show the inferior surface.

Family XLI. ALAUDINÆ.

Genus *Alauda.* LINNÆUS.

Alauda arvensis. L. **The Skylark.**

474. Skeleton. Presented by Prof. Clark.
475. Head,   do.
476. Head reversed, to show the inferior surface.
Presented by Prof. Clark.

Family XLV. LIOTRICHINÆ.

Genus *Mimus.* BOIE.

Mimus polyglottus. L. **The Mocking-bird.**

477. Skeleton. From the Zoological Society's Gardens.
Presented by Prof. Clark.

Family XLVI. TURDINÆ.

Genus *Turdus.* LINNÆUS.

Turdus merula. L. **The Blackbird.**

478. Skeleton. Presented by Prof. Clark.
479. Head.   do.
480. Head reversed, to show the inferior surface.
Presented by Prof. Clark.
481. Sternum. Presented by Prof. Clark.
482. Pelvis.   do.

Turdus macrourus. GMEL.

483. Stuffed specimen. Presented by Thos. Horsfield, M.D.

6

Family XLVII. MOTACILLINÆ.

Genus *Sylvia*. LATHAM.

Sylvia rubecula. LATH. **The Robin Red-breast.**

484. Head. Presented by Prof. Clark.

Sylvia hortensis. LATH. **The Garden Warbler.**

485. Head. Presented by Prof. Clark.
486. Sternum, coracoid bones, clavicle, scapulæ and wings.

Presented by Prof. Clark.

Family XLVIII. MUSCICAPA.

Genus *Muscicapa*. LINNÆUS.

Muscicapa Sparmanni.

487. Stuffed specimen. Presented by Thomas Horsfield, M.D.

Family XLIX. LANIINÆ.

Genus *Edolius*. CUVIER.

Edolius remifer. TEMM.

488. Stuffed specimen. Presented by Thomas Horsfield, M.D.

Family L. CHELIDONES.

Genus *Hirundo*. LINNÆUS.

Hirundo riparia. L. **The Sand Martin.**

489. Head. Presented by Prof. Clark.

Hirundo urbica. L. **The House Martin.**

490. Head. Presented by Prof. Clark.

Hirundo rustica. L. **The Swallow.**

491. Skeleton. Presented by Prof. Clark.
492. Head.   do.

Family LI. NYCTICHELIDONES.

Genus *Podargus*. CUVIER.

Podargus phalænoides. GOULD.

493. Head, imperfect. Presented by Prof. Clark.

Order VI. RAPTATORES.

Family LII. STRIGIDÆ.

Genus *Strix*. SAVIGNY.

Strix flammea. L. **The Barndoor Owl.**

494. Skeleton. Macartney collection.
495. Skeleton. Presented by Prof. Clark.
496. Head. do.

Strix Ceylonensis. GMEL.

497. Stuffed specimen. Presented by Thomas Horsfield, M.D.

Family LIII. ACCIPITRINÆ.

Genus *Falco*. LINNÆUS.

Falco peregrinus. GMEL. **Peregrine Falcon.**

498. Skeleton. Presented by Prof. Clark.
499. Sternum, coracoid bones, clavicle and scapulæ.
Presented by Prof. Clark.

Falco tinnunculus. L. **The Kestrel.**

500. Skeleton. Presented by Prof. Clark.

Falco Nisus. L. **The Sparrow-Hawk.**

501. Skeleton. Presented by Prof. Clark.
502. Head. do.

Falco Pondicerianus. GMEL.

503. Stuffed specimen. Presented by Thomas Horsfield, M.D.

Falco Bacha. DAUD.

504. Stuffed specimen. Presented by Thomas Horsfield, M.D.

Genus *Milvus.* BECHSTEIN.

Milvus regalis. BRISSON. **The Kite.**

505. Sternum, coracoid bones, clavicle and scapulæ.
Presented by Prof. Clark.

Genus *Buteo.* BECHSTEIN.

Buteo vulgaris. BECHST. **The Common Buzzard.**

506. Head. Presented by H. J. H. Bond, M.D. Corpus Christi Coll.
507. Head reversed, to show the inferior surface. do.

Genus *Aquila.* BRISSON.

Aquila chrysaëtos. CUVIER. **The Golden Eagle.**

508. Skeleton of a male. Macartney collection.
509. Skeleton, placed in the act of expanding its wings, in order to show the mode in which the ligaments that attach the quills to the bones of the forearm are tightened when the limb is extended.
Brookes collection.

Aquila (Species uncertain).

510. Left tarsus and toes. Presented by Prof. Clark.
511. Right tarsus and toes. do.

Genus *Haliaëtus.* SAVIGNY.

Haliaëtus albicilla. SAV. **The White-tailed Eagle.**

512. Skeleton. Brookes collection.
513. Head from Iceland. The bird was found drowned in the nets of some fishermen, in which it is supposed it was caught while fishing. Presented by J. W. Clark, M.A.
514. Sternum.
Presented by Rev. W. T. Kingsley, M.A. Sidney College.

Genus *Pandion.* SAVIGNY.

Pandion haliaëtus. SAV. **The Osprey.**

515. Sternum, coracoid bones, clavicle and scapulæ.

Presented by Prof. Clark.

Genus *Gypogeranus.* ILLIGER.

Gypogeranus serpentarius. L. **The Cape Sagittary.**

516. Skeleton. Brookes collection.

Family LIV. VULTURINÆ.

Genus *Vultur.* LINNÆUS.

Vultur (Species uncertain.)

517. Head. Harwood collection.

Genus *Cathartes.* ILLIGER.

Cathartes gryphus. ILL. **The Condor.**

518. Head. [French Expedition.] Presented by Prof. Clark.
519. Left foot, dried. Harwood collection.

CLASS IV. MAMMALS. *(Mammalia.)*

Order I. MONOTREMATA.

Family I. MONOTREMATA.

Genus *Ornithorynchus.* BLUMENBACH.

Ornithorynchus paradoxus. BLUMENB. **The Duck-billed Animal: the Water-mole.**

Cervical vertebræ...		7.
Dorsal	do.	... 17.
Lumbar	do.	... 2.
Sacral	do.	... 2.
Caudal	do.	... 20.

520. Skeleton, prepared by Sir Everard Home. Brookes collection.

521.   Head, with the integuments on the upper and lower jaw preserved in their natural state. Brookes collection.

522.   A stuffed specimen. Macartney collection.

Genus *Echidna*. CUVIER.

Echidna setosa. CUV. **The Short-spined Echidna.**

Cervical vertebræ... 7.
Dorsal   do.   ... 16.
Lumbar   do.   ... 3.
Sacral   do.   ... 3.
Caudal   do.   ... 2 (the rest are wanting).

523.   Skeleton. Presented by the Cambridge Philosophical Society.

Order II. MARSUPIALIA.

Family II. GLIRINA.

Genus *Phascolomys*. GEOFFROY ST HILAIRE.

Phascolomys Wombat. PÉRON. **The Wombat.**

524.   Head. Presented by the Cambridge Philosophical Society.

Family III. MACROPODA.

Genus *Macropus*. SHAW.

Macropus giganteus. SHAW. **The Great Kangaroo.**

Cervical vertebræ... 7.
Dorsal   do.   ... 13.
Lumbar   do.   ... 6.
Sacral   do.   ... 2.
Caudal   do.   ... 20.

525.   Skeleton. [Bell collection.] Presented by Prof. Clark.

526.   Head.

527.   Vertebral column, with the ribs and pelvis; (there are twenty-two caudal vertebræ). } From an imperfect skeleton. Macartney collection.

528.   Right hind leg.

529.   Left hind leg.

530.   Head. [Bell collection.] Presented by Prof. Clark.

531.   Right hand, dried. Harwood collection.

532. Right hind foot, dried. Presented by Prof. Clark.

533. Right hind foot, dried.} Presented by Prof. Clark.
534. Left hind foot, dried. }

Macropus (Species uncertain).

535. Bones of a young animal, incomplete. Macartney collection.

Cervical vertebræ ... 7.
Dorsal do. ... 13.
Lumbar do. ... 6.
Sacral do. ... 2.
Caudal do. ... 13 (the rest are wanting).

536. Head of a small species. [Bell collection.]
Presented by Prof. Clark.

Genus *Hypsiprymnus.* ILLIGER.

Hysiprymnus murinus. ILLIG. **The Potoroo, or Rat-Kangaroo.**

Cervical vertebræ ... 7.
Dorsal do. ... 13.
Lumbar do. ... 6.
Sacral do. ... 2.
Caudal do. ... 25.

537. Skeleton. [Bell collection.] Presented by Prof. Clark.

Hypsiprymnus penicillatus. OGILBY. **Tufted-tailed Rat-Kangaroo.**

538. Head. Brookes collection.

Family IV. PHALANGISTÆ.

Genus *Phalangista.* CUV.

Phalangista vulpina. DESMAR.

Cervical vertebræ ... 7.
Dorsal do. ... 13.
Lumbar do. ... 6.
Sacral do. ... 2.
Caudal do. ... 25.

539. Skeleton. [Bell collection.] Presented by Prof. Clark.

48 MAMMALS.

Genus *Petaurus*. SHAW.

Petaurus Ariel. WATERHOUSE.

540. Head. Brookes collection.

Family VII. PEDIMANA.

Genus *Didelphis*. LINNÆUS.

Didelphis virginiana. SHAW. **The Virginian Opossum.**

Cervical vertebræ ... 6 (the atlas is wanting).
Dorsal   do.   ... 13.
Lumbar   do.   ... 6.
Sacral   do.   ... 2.
Caudal   do.   ... 21 (several wanting).

541. Skeleton. Brookes collection.

Didelphis (Species uncertain).

542. Head. [Bell collection.] Presented by Prof. Clark.
543. Head. Brookes collection.

Genus *Diprotodon*. OWEN. (*Fossil.*)

544. A plaster cast of a portion of the right ramus of the lower
jaw. Presented by the Royal College of Surgeons of England.

Order III. CETACEA.

Family VIII. CETACEA.

Genus *Balæna*. LINNÆUS. **The True Whales.**

Balæna mysticetus. L. **The Whalebone Whale.**

545. The tympanic bone of the right side.⎫ [French Exp.]
546. The tympanic bone of the left side. ⎭ Presented by Prof. Clark.

Genus *Balænoptera*. LACEPEDE. **The Finner Whales.**

Balænoptera rostrata. LAC. **The Pike Whale.**

Cervical vertebræ ... 7.
Dorsal   do.   ... 11.
Caudal   do.   ... 34.

547. Skeleton. Purchased by subscription.

Genus *Physeter*. LINNÆUS.

Physeter macrocephalus. SHAW. **The Spermaceti Whale, or Cachalot.**

548. The lower jaw. There are twenty-two pairs of teeth: the symphysis of the jaw begins opposite the eighteenth pair. In Owen's specimen of a female there are also twenty-two teeth on each side: but in that of a male twenty-seven. From the Museum of J. P. Delafons, Esq. Presented by Prof. Clark.

The following specimens are of uncertain species:

549. The tympanum of a Cetacean. [Bell collection.]
Presented by Prof. Clark.

550. Vertebra towards the end of the tail of an enormous Cetacean. [Bell collection.] Presented by Prof. Clark.

551. Vertebra towards the middle of the back of a much smaller Cetacean. [Bell collection.] Presented by Prof. Clark.

552. Body of a vertebra of a Cetacean, with the two epiphyses detached. Presented by Prof. Henslow.

553. Body of a vertebra of a Cetacean, found in excavating a well in Norfolk, at a depth of seventy-two feet.
Presented by Mr Lichfield, Cambridge.

554. Portion of a rib of an enormous Cetacean, from the right side. [Bell collection.] Presented by Prof. Clark.

Genus *Ziphius*. CUVIER.

Ziphius Sowerbiensis. GRAY.

555. A plaster cast of the head. Teeth two, large, compressed, in the lower jaw. Presented by Prof. Acland, of Oxford.

Genus *Delphinus*. LINNÆUS.

Delphinus Orca. L. **The Grampus.**

556. Skeleton. Number of alveoli: $\frac{11-11}{12-12} = 46$. OWEN. In this specimen the number is $\frac{12-12}{13-13} = 50$.

7

There are seven cervical vertebræ, of which the first four coalesce. There are seven true ribs, in all twelve: the first seven attached by their heads and their tubercles: the others by their transverse processes only. The sternal portion of these ribs is, osseous. There are twelve vertebræ carrying ribs (dorsal vertebræ): thirty-three lumbar or caudal. The inferior spines or chevron bones first appear on the eleventh of these vertebræ, and cease about the twenty-second or twenty-third. Macartney collection.

557.   Head. Number of alveoli: $\frac{12-12}{11-11}=46$. A number which suits the dentition of the Black Grampus (*Delphinus melas*). Macartney collection.

Delphinus phocæna. L. **The Porpoise.**

Cervical vertebræ ... 7 (anchylosed).
Dorsal   do.   ... 12.
Caudal   do.   ... 46.

558.   Skeleton, wanting the arms below the scapulæ. Teeth are in part deficient. Harwood collection.

559.   Bones of a young skeleton, complete.

Presented by Prof. Clark.

560.   Head, wanting the teeth and the lower jaw, of an old individual. Alveoli, 18 or 19, on each side, somewhat indistinct. It was found on the Holderness coast, by the Rev. Christopher Sykes, and presented by him to the Museum.

561.   Head, with the tympanic bones *in situ.* Number of teeth: $\frac{29-29}{25-25}=108$. Macartney collection.

562.   Forty-two consecutive vertebræ, of the same animal.

Macartney collection.

563.   Right anterior extremity. ⎫ From the coast of Belgium.
564.   Left anterior extremity. ⎬   Presented by Prof. Clark.

Delphinus Tursio. Fabr. **The Bottle-nosed Whale.**

565.   Head. Teeth: $\frac{24-23}{25-25}=97$. [Bell collection.] Presented by Prof. Clark.

566. Head. Teeth: $\frac{21-21}{22-22}=86$. The number varies in different specimens. [Bell collection.] Presented by Prof. Clark.

Delphinus leucas. PALL. **The Beluga.**

567. Head. Teeth: $\frac{10-10}{9-9}=38$. [Bell collection.]
Presented by Prof. Clark.

568. Head. Teeth: $\frac{8-9}{6-7}=30$. Owen gives the number $\frac{9-9}{9-9}=36$. [Bell collection.] Presented by Prof. Clark.

Delphinus longirostris. GRAY. **The Cape Dolphin.**

569. Head. Teeth: $\frac{48-48}{45-45}=186$. [French Exp.] From Torres Straits. Presented by Prof. Clark.

Delphinus delphis. L. **The Dolphin.**

570. Head. Teeth: $\frac{48-46}{45-45}=184$. [Bell collection.]
Presented by Prof. Clark.

571. Head. Teeth: $\frac{47-47}{47-48}=189$. [Bell collection.]
Presented by Prof. Clark.

572. Head. [Bell collection.] Presented by Prof. Clark.

Delphinus (Species uncertain).

Cervical vertebræ ... 7.
Dorsal   do.   ... 11.
Caudal   do.   ... 30.

573. Skeleton, wanting the head and the anterior extremities, of a large species, perhaps of a Narwhal (*Monodon monoceros.* L.). The cervical vertebræ are all distinct, and the first and second of great size. Harwood collection.

574. Right scapula, of the same.

575. The left arm, of the same.

576. A dorsal vertebra, of the same.

577. Fragment of a cranium, from the Straits of Magellan. [French Exp.] Presented by Prof. Clark.

578. Lower jaw of a young specimen, akin to *D. phocœna.* Teeth: 41 – 44. [Bell collection.] Presented by Prof. Clark.

Family IX. SIRENIA.

Genus *Halicore.* ILLIGER.

Halicore Australis. OWEN. **The Australian Dugong.**

Dental formula: $i. \frac{1-1}{4-4}, \ m. \frac{4-4}{5-5}.$

579. Head, from Torres Straits. [French Exp.]
Presented by Prof. Clark.

580. Head, from the same locality. Presented by Prof. Clark.

581. Fragment of a skull, from the same locality.
Presented by Prof. Clark.

582. Fragment of a skull, from the same locality.
Presented by Prof. Clark.

583. An incisor tooth of a very old individual.
Presented by Prof. Clark.

Order IV. PACHYDERMATA.

Family X. ELEPHANTINA.

Genus *Mastodon.* CUVIER. (*Fossil.*)

Mastodon giganteus. CUV.

584. A molar tooth. Presented by Prof. Clark.

585. Portion of a molar tooth. [Bell collection.]
Presented by Prof. Clark.

586. Cast of a molar tooth. [Bell collection.]
Presented by Prof. Clark.

587. Cast of a molar tooth. [Bell collection.]
Presented by Prof. Clark.

Genus *Elephas*. LINNÆUS.

Elephas indicus. Cuv. **The Indian Elephant.**

Dental formula: $i. \frac{2-2}{0-0}$, $m. \frac{6-6}{6-6} = 28$, appearing in succession from behind forward.

Cervical vertebræ ...		7.
Dorsal	do. ...	19.
Lumbar	do. ...	4.
Sacral	do. ...	3.
Caudal	do. ...	30.

588. Skeleton. Presented by Prof. Clark.

589. Head, with the mouth open to show the teeth. Macartney collection.

590. Head. A section has been carried perpendicularly behind the petrous bones to show the cavity of the cranium, and the extensive system of cells in connection with the mastoid cells. Macartney collection.

591. The occipital portion of the same skull.

592. Left half of the lower jaw. The section shows the second molar advancing upon the first. [Bell collection.]
Presented by Prof. Clark.

593. A transverse and perpendicular section of a molar tooth.
Presented by Prof. Clark.

594. The second molar tooth of the upper jaw, left side. [Bell collection.] Presented by Prof. Clark.

595. The second molar of the upper jaw, left side, just beginning to be worn. Presented by James Hildyard, M.A. Christ's College.

596. The left incisor tooth. [Bell collection.]
Presented by Prof. Clark.

597. Right femur, divided longitudinally, showing that there is no medullary cavity. Presented by Prof. Clark.

598. The sabot of the right fore foot.
599. The sabot of the left fore foot. } Brookes collection.

The following bones are from a young elephant. Brookes collection:

600.   Right forearm.

601.   Right humerus.

602.   Left humerus.

603.   Left radius and ulna.

604.   Right femur.

605.   Left femur.

606.   Right hind leg.

607.   Left hind leg.

608.   Second and six following vertebræ, showing their original elements.

609.   Four dorsal vertebræ in a similar state.

610.   Two sacral, with the first six caudal vertebræ.

Elephas africanus. Cuv. **The African Elephant.**

611.   A transverse and perpendicular section of a portion of a molar tooth. [Bell collection.] Presented by Prof. Clark.

612.   Cast of the anterior tooth of the lower jaw, left side. [Bell collection.] Presented by Prof. Clark.

Elephas. Bones from fossil specimens.

613.   Left femur, from the gravel at Barnwell Abbey. Found at a depth of fourteen feet, in many fragments.

Presented by the devisees of the late Chas. Geldart, LL.D.

614.   Portion of the pubic bone, of the left side. From the same locality.

Presented by J. C. Geldart, LL.D. Master of Trinity Hall.

615.   A fragment of the iliac bone, of the left side, exhibiting a portion of the acetabulum. From the same locality.

Presented by J. C. Geldart, LL.D. Master of Trinity Hall.

616.   Portion of the ischium, of the left side. From the same locality.

Presented by J. C. Geldart, LL.D. Master of Trinity Hall.

617.   Third metatarsal bone, left side. From the same locality.

Presented by Prof. Clark.

618. Two sections of the tusk of an elephant, showing the disposition to separate into concentric laminæ. [Bell collection.]

Presented by Prof. Clark.

619. A molar tooth. [Bell collection.]

Presented by Prof. Clark.

620. Portions of a fossil tooth, splitting into its component denticles.

Presented by the devisees of the late Chas. Geldart, LL.D.

621. Second molar tooth, upper jaw, left side. [Bell collection.]

Presented by Prof. Clark.

622. Second molar tooth, lower jaw, left side. [Bell collection.]

Presented by Prof. Clark.

623. A lamina of a tooth. From Kent's cavern, Torquay, Devon. Macartney collection.

624. Cast of the right femur. The original is preserved in Jesus College. Locality unknown.

Presented by J. Okes, Esq. Sidney College.

Family XI. NASICORNIA.

Genus *Rhinoceros*. LINNÆUS.

Dental formula: $i. \frac{1-1}{2-2}, \quad p. \frac{4-4}{4-4}, \quad m. \frac{3-3}{3-3} = 34.$

Rhinoceros indicus. Cuv. **The Indian Rhinoceros.**

Cervical vertebræ ... 7.
Dorsal   do.   ... 19.
Lumbar   do.   ... 3.
Sacral   do.   ... 4.
Caudal   do.   ... 22 (a few are wanting).

625. Skeleton, the bones from Wombwell's Menagerie, of a young individual. The dentition is proceeding.

Presented by Prof. Clark.

626. Humerus, with the radius and ulna articulated, of the right side. Brookes collection.

627. Left humerus. Brookes collection.

628. Left femur.   do.

629. Left radius.   do.

630. Left ulna.   do.

631.   The horn. Macartney collection.

632.   The horn, resolved at its base into its component hairs. Macartney collection.

Rhinoceros tichorinus. Cuvier. (*Fossil.*)

Fossil bones from Kent's cavern. Macartney collection.

633.   An upper molar tooth, right side.

634.   An unworn upper molar, right side.

635.   First premolar, left side, of a much smaller animal.

636.   A lower molar.

637.   A lower molar.

638.   A lower molar.

639.   Portion of the acetabulum.

640.   Portion of the left tibia.

641.   Fragment of a humerus.

642.   Portion of a femur.

Family XII. Lamnungia.

Genus *Hyrax.* Cuvier.

Dental formula: $i.\ \dfrac{1-1}{2-2},\ p.\ \dfrac{4-4}{4-4},\ m.\ \dfrac{3-3}{3-3} = 34.$

Hyrax capensis. Schreber. **The Klip-das.**

Cervical vertebræ ...   7.
Dorsal   do.   ... 21.
Lumbar   do.   ... 8.
Sacral   do.   ... 2.
Caudal   do.   ... 5 (the rest are wanting).

643.   Skeleton. [Bell collection.] Presented by Prof. Clark.

Family XIII. Tapirina.

Genus *Tapirus.* Brisson.

Dental formula: $i.\ \dfrac{3-3}{3-3},\ c.\ \dfrac{1-1}{1-1},\ p.\ \dfrac{4-4}{3-3},\ m.\ \dfrac{3-3}{3-3} = 42.$

Tapirus americanus. Auct. **The American Tapir.**

644.   Head. [Bell collection.] Presented by Prof. Clark.

Family XIV. SOLIDUNGULA.

Genus *Equus*. LINNÆUS.

Dental formula: $i.\ \frac{3-3}{3-3},\ c.\ \frac{1-1}{1-1},\ p.\ \frac{3-3}{3-3},\ m.\ \frac{3-3}{3-3}=40.$

Equus caballus. L. **The Horse.**

Cervical vertebræ ... 7.
Dorsal do. ... 17.
Lumbar do. ... 6 (the last two partially anchylosed).
Sacral do. ... 4.
Caudal do. ... 16 (the rest wanting).

645. Skeleton of a thorough-bred horse, four years old.

Presented by Prof. Clark.

At this age there are six lumbar vertebræ; in the adult state the sixth is anchylosed to the fifth. At this period there are two epiphyses between the bodies of the fifth and sixth, but the transverse processes are already anchylosed beyond the sacral foramina. In the sacrum the first vertebra alone is articulated to the *Os innominatum.*

646. Head. Presented by J. Okes, Esq. Sidney College.

647. Head. Presented by Prof. Clark.

648. The fifth cervical vertebra. [Bell collection.]

Presented by Prof. Clark.

649. Left humerus. Harwood collection.

650. Longitudinal section of the left humerus, in front of the head, to show the reticular structure of the upper part, and the medullary canal in the middle of the shaft. Harwood collection.

651. Left tibia of a racer, broken whilst exercising at Newmarket.

Presented by Wm. Peck, Esq.

The separate bones of the right hind leg of a cart horse.

Presented by Prof. Clark.

652. Tibia, or *leg-bone.*

The tarsus or *hock,* consisting of

653. The astragalus.

654. The os calcis.

8

655. The cuboid bone.

656. The cuneiform bones.

657. The metatarsal or *cannon-bone*, with the two lateral incomplete metatarsal bones, called *splint-bones*, and the sesamoid bones.

658. The first phalanx, or *great pastern*.

659. The second phalanx, or *little pastern*.

660. The third phalanx, or *coffin-bone*, with its sesamoid bone, commonly called the *navicular*, or *shuttle-bone*.

661. The right hoof. Presented by Prof. Clark.

662. A monstrous foal, curious from the imperfect ossification of the scapulæ, showing the acromion process formed of a distinct piece. Presented by Prof. Clark.

Fossil bones from the gravel, near Cambridge:

663. Right femur.

664. The sacrum.

665. Portion of the sacrum of a smaller individual, consisting of portions of four sacral vertebræ.

666. Inferior extremity of the right radius.

667. Part of the right metatarsal bone.

Fossil teeth from Kent's Cavern. Macartney collection.

668. Portion of a molar tooth imbedded in diluvium.

669. An incisor tooth of the lower jaw, right side.

670. The last true molar of the lower jaw, left side.

671. A molar of the lower jaw, left side.

672. The last true molar of the upper jaw, left side.

673. A molar of the upper jaw, right side.

Genus *Macrauchenia*. OWEN. (*Fossil.*)

Macrauchenia Patachonica. OWEN.

674. Casts of three metacarpal bones of the right forefoot.
　　　　Presented by the Royal College of Surgeons of England.

675. A cast of the right femur.
　　　　Presented by the Royal College of Surgeons of England.

676. A cast of the fourth or fifth cervical vertebra.

Presented by the Royal College of Surgeons of England.

Equus asinus. LINNÆUS. **The Ass.**

Cervical vertebræ ... 7.
Dorsal   do.   ... 18.
Lumbar   do.   ... 5.
Sacral   do.   ... 5.
Caudal   do.   ... 3 (the rest are wanting).

677. Skeleton. Harwood collection.

678. Head. The external lamina of bone has been removed from the jaw on the left side, to show the dentition. [Bell collection.]
Presented by Prof. Clark.

679. The upper jaw, with the outer lamina of the superior maxillary bone removed to show the dentition of the molar teeth. The three permanent molars are about to descend. The premolars are causing absorption of the roots of the deciduous molars.
Presented by Prof. Clark.

680. The lower jaw, with the lamina removed on the inside. The process of dentition is shown, corresponding with that of the upper jaw. Presented by Prof. Clark.

681. Right humerus. Presented by Prof. Clark.

682. A longitudinal section of the right humerus. Harwood collection.

683. Right radius and ulna. Presented by Prof. Clark.

684. A portion of the carpus, metacarpal bones, phalanges, and hoof of the right fore leg. Harwood collection.

685. The carpus, metacarpal bones, and phalanges of the left fore leg. Harwood collection.

686. The tarsus, metatarsal bones, and phalanges of the left hind leg, with the tendons of the muscles: the arteries injected. Harwood collection.

687. The five lumbar vertebræ of a young ass.
Presented by Prof. Clark.

688.   Four sacral vertebræ, partially anchylosed, of the same.

Presented by Prof. Clark.

689.   Right astragalus, of the same.   Presented by Prof. Clark.

690.   Left astragalus, of the same.   do.

Family XV. SUINA.

Genus *Phacochœrus.* CUV.

Dental formula: $i.\ \dfrac{1-1}{3-3},\ c.\ \dfrac{1-1}{1-1},\ p.\ \dfrac{2-2}{2-2},\ m.\ \dfrac{3-3}{3-3}=32.$

Phacochœrus Æliani. RUEPPEL. **The African Wart-hog.**

691.   Head.   Presented by Prof. Clark.

Genus *Sus.* LINNÆUS.

Dental formula: $i.\ \dfrac{3-3}{3-3},\ c.\ \dfrac{1-1}{1-1},\ p.\ \dfrac{4-4}{4-4},\ m.\ \dfrac{3-3}{3-3}=44.$

Sus scrofa. L. **The Common Pig.**

Cervical vertebræ ...   7.
Dorsal   do.   ... 14.
Lumbar   do.   ...   5.
Sacral   do.   ...   5.
Caudal   do.   ...   4 (the rest are wanting).

692.   Skeleton.   Presented by G. M. Humphry, M.D.

693.   Skeleton of a large boar pig.   The bones presented by Mr Hopkins, Brewer, Cambridge.

694.   Head of a wild boar.   [Bell collection.]

Presented by Prof. Clark.

695.   Head.   [Bell collection.]   Presented by Prof. Clark.

696.   Head of a boar pig.

Presented by Mr Hopkins, Brewer, Cambridge.

697.   Head of a boar pig, diseased.

Presented by Mr Hopkins, Brewer, Cambridge.

698.   Head.   From India.   Presented by Huddlestone Stokes, Esq.

699. Head of an Indian wild boar.
>> Presented by Huddlestone Stokes, Esq.

700. Head of a variety from St Croix, Teneriffe. [French Exp.]
>> Presented by Prof. Clark.

701. Head of a variety from Ceram I. [French Exp.]
>> Presented by Prof. Clark.

702. Head of a variety from Balambangan, Borneo. [French Exp.]
>> Presented by Prof. Clark.

703. Head of a variety from Poverty Bay, New Zealand. [French Exp.] Presented by Prof. Clark.

Sus Babyrussa. L. **The Babyrussa,** or **Stag-hog.**

704. Head. Harwood collection.
705. Head. [Bell collection.] Presented by Prof. Clark.
706. Head. do. do.

Family XVI. HIPPOPOTAMINA.

Genus *Hippopotamus.* L.

Dental formula: $i. \frac{2-2}{2-2},\ c. \frac{1-1}{1-1},\ p. \frac{4-4}{4-4},\ m. \frac{3-3}{3-3}=40.$

Hippopotamus amphibius. L. **The Hippopotamus.**

Cervical vertebræ ... 7.
Dorsal do. ... 15.
Lumbar do. ... 4.
Sacral do. ... 5.
Caudal do. ... 16.

707. Skeleton. From the Museum of J. P. Delafons, Esq.
>> Presented by Prof. Clark.

708. Head. [Bell collection.] Presented by Prof. Clark.
709. The sabot of the right forefoot. Brookes collection.
710. The sabot of the left forefoot. Brookes collection.

Order V. RUMINANTIA.

Family XVII. TYLOPODA.

Genus *Camelus*. L.

Dental formula: $i. \frac{1-1}{3-3}$, $c. \frac{1-1}{1-1}$, $p. \frac{3-3}{2-2}$, $m. \frac{3-3}{3-3} = 34$.

Camelus dromedarius. L. **The Camel.**

Cervical vertebræ ... 7.
Dorsal do. ... 12.
Lumbar do. ... 7.
Sacral do. ... 4.
Caudal do. ... 10 (the rest are wanting).

711. Skeleton of a young female, from Bell's travelling menagerie.
             Presented by Prof. Clark.

712. Head. Presented by Prof. Clark.

713. Lower jaw, picked up near the pyramids of Cairo, and presented by John Anthony, M.D. Caius College.

Genus *Auchenia*. ILLIGER.

Dental formula: $i. \frac{1-1}{3-3}$, $c. \frac{1-1}{1-1}$, $p. \frac{2-2}{1-1}$, $m. \frac{3-3}{3-3} = 30$.

Auchenia lama. BRANDT. **The Llama, or Peruvian Camel.**

Cervical vertebræ ... 7.
Dorsal do. ... 12.
Lumbar do. ... 7.
Sacral do. ... 5.
Caudal do. ... 9 (the rest are wanting).

714. Skeleton. Macartney collection.

715. Bones of a skeleton, complete, from Wombwell's menagerie.
             Presented by Prof. Clark.

716. Head of a male, from Patagonia. [French Exp.]
             Presented by Prof. Clark.

717. Head, wanting the lower jaw. [French Exp.]
Presented by Prof. Clark.

718. Right ramus of the lower jaw of an older individual. [French Exp.] Presented by Prof. Clark.

Family XVIII. ELAPHII.

Genus *Moschus*. L.

Dental formula: $i.\ \frac{0-0}{4-4},\ c.\ \frac{1-1}{0-0},\ p.\ \frac{3-3}{3-3},\ m.\ \frac{3-3}{3-3}=34.$

Moschus Javanicus. GMEL. The Javan Chevrotain.

Cervical vertebræ ... 7.
Dorsal   do.   ... 13.
Lumbar   do.   ... 6.
Sacral   do.   ... 5.
Caudal   do.   ... 8.

719. Skeleton of a female. From the Zoological Society's gardens.
Presented by Prof. Clark.

720. A stuffed specimen. Presented by Thomas Horsfield, M.D.

Moschus pygmæus. LINNÆUS. The Pigmy Chevrotain.

Cervical vertebræ ... 7.
Dorsal   do.   ... 13.
Lumbar   do.   ... 6 (these are succeeded by fifteen vertebræ, none of which are as yet anchylosed to form a sacrum).

721. Skeleton. [Bell collection.] Presented by Prof. Clark.

722. Skeleton, incomplete. There are four sacral vertebræ. [Bell collection.] Presented by Prof. Clark.

Moschus moschiferus. L.

723. The muzzle, with long canine teeth. Probably of a male. Harwood collection.

Moschus (Species uncertain).

724.   Head, wanting the lower jaw.   [Bell collection.]
                                    Presented by Prof. Clark.

725.   Head, with the left ramus of the lower jaw.   [Bell collection.]
                                    Presented by Prof. Clark.

Genus *Cervus.* L.

Dental formula: $i.\ \dfrac{0-0}{4-4},\quad c.\dfrac{1-1}{0-0},\quad p.\dfrac{3-3}{3-3},\quad m.\dfrac{3-3}{3-3}=34.$

Cervus Alces. L. **The Elk.**

726.   Head and horns. [Bell collection.] Presented by Prof. Clark.

Cervus Tarandus. L. **The Reindeer.**

Cervical vertebræ ...	7.
Dorsal do. ...	14.
Lumbar do. ...	5.
Sacral do. ...	5.
Caudal do. ...	9.

727.   Skeleton of a female.   Macartney collection.

728.   Bones of a young male.   There are only four sacral vertebræ.
[Bell collection.]   Presented by Prof. Clark.

729.   Head, imperfect, showing the origin of the horns.   [Bell collection.]   Presented by Prof. Clark.

730.   Head, with a variety in the branching of the horns, especially
on the left side.   From Hammerfest, Norway.
                          Presented by J. W. Clark, M.A. Trin. Coll.

731.   Head and horns of an older individual.   From the same
locality.   Presented by J. W. Clark, M.A. Trin. Coll.

732.   Horns of a male, of great size.   From Throndhjem, Norway.
                          Presented by J. W. Clark, M.A. Trin. Coll.

733.   Lower jaw.   [Bell collection.]   Presented by Prof. Clark.

Cervus elaphus. L. **The Red Deer.**

734.   Skeleton of a female.   Presented by Prof. Clark.

Cervus Axis. L.

735. Head of a male, with its horns. From Mindanao Island. [French Exp.] Presented by Prof. Clark.

736. Head of a male, with its horns, dried. [Bell collection.]
Presented by Prof. Clark.

737. Horns of a large male. Presented by Huddlestone Stokes, Esq.

Cervus Muntjak. ZIMMERMAN. **The Barking Deer.**

738. Head, wanting the lower jaw. Harwood collection.

739. Head, complete.
Presented by Mr Baker, Naturalist, Cambridge.

Cervus Dama. L. **The Fallow Deer.**

Dental formula: $i. \frac{0-0}{4-4}, \quad c. \frac{0-0}{0-0}, \quad p. \frac{3-3}{3-3}, \quad m. \frac{3-3}{3-3} = 32.$

Cervical vertebræ ... 7.
Dorsal   do.   ... 13.
Lumbar   do.   ... 6.
Sacral   do.   ... 4.
Caudal   do.   ... 11.

740. Skeleton of a male. Presented by Prof. Clark.

741. Skeleton of a female. There are five sacral vertebræ. The last appears to have been anchylosed subsequent to the first four.
Presented by Prof. Clark.

742. Bones of a female. There are four sacral vertebræ.
Presented by Prof. Clark.

743. Bones of a young female. There are four sacral vertebræ.
Presented by Earl Fitzwilliam.

744. Head of a male, with its horns, wanting the lower jaw. [Bell collection.] Presented by Prof. Clark.

745. Right half of the skull of a male, with its horn. [Bell collection.] Presented by Prof. Clark.

746. Left half of the same skull.

747.   Right horn in a state of growth. [Bell collection.]
                                     Presented by Prof. Clark.

748.   Right horn in a state of growth. [Bell collection.]
                                     Presented by Prof. Clark.

749.   Left horn in a state of growth. [Bell collection.]
                                     Presented by Prof. Clark.

Cervus capreolus. L. **The Roe-deer.**

750.   Frontlet and antlers. [Bell collection.]
                                     Presented by Prof. Clark.

751.   Right horn. [Bell collection.] Presented by Prof. Clark.

The following bones of a large deer, probably *Cervus elaphus*, were presented by Mr Clark of Saffron Walden:

752.   Right tibia.
753.   Left tibia.
754.   Right radius.
755.   Left radius.
756.   Right ulna.
757.   Left ulna.
758.   Right metacarpal bone.
759.   Left metacarpal bone.
760.   Right metacarpal bone, of another animal.

Cervus giganteus. BLUMENBACH. (*Fossil.*)

Megaceros hibernicus. OWEN. **The Gigantic Irish Deer,** commonly called **The Irish Elk.**

761.   Head and horns.

762.   The second, third, fourth, fifth, and sixth dorsal vertebræ, with some of the ribs, or portions of ribs, attached.

763.   Left scapula.

764.   Right foreleg, complete with the exception of the pisiform bone.

765. Tibia, tarsus, metatarsal bone, three phalanges of the inner toe, and one of the outer, of the left hind leg.

766. Right half of the lower jaw of another specimen.

767. Left half of the lower jaw.

768. A dorsal vertebra, probably the second or third.

769. A tray of fragments of bones.

The above bones were found near Armagh, Ireland, and presented by Rev. Richard Allott, M.A. Trin. Coll.

770. An upper molar tooth, right side. From Kent's Cavern. Macartney collection.

Genus *Strongyloceros.* OWEN.

Strongyloceros spelæus. OWEN. (*Fossil.*)

771. Fragment of the base of an antler. From Kent's Cavern. Macartney collection.

772. A lower molar tooth, right side, of an extinct deer, from Kent's Cavern. Macartney collection.

Genus *Poebothrium.* LEIDY.

Poebothrium Wilsonii. LEIDY.

773. Cast of the head.

Presented by the Royal College of Surgeons of England.

Genus *Camelopardalis.* GMELIN.

Dental formula: $i. \frac{0-0}{4-4}, \ c. \frac{0-0}{0-0}, \ p. \frac{3-3}{3-3}, \ m. \frac{3-3}{3-3} = 32.$

Camelopardalis Giraffa. GMEL. **The Giraffe.**

Cervical vertebræ ... 7.
Dorsal do. ... 13.
Lumbar do. ... 6.
Sacral do. ... 3 (another would probably have become anchylosed shortly).
Caudal do. ... 21.

774. Young skeleton. The animal died in Wombwell's menagerie, at Norwich. Presented by Professor Clark.

Family XIX. CAVICORNIA.

Genus *Antilope*. PALLAS.

Dental formula the same as *Camelopardalis.*

Antilope strepsiceros. PALL. The Koodoo, or Striped Antelope.

775. Head, with the horns, wanting the lower jaw. [Bell collection.] Presented by Prof. Clark.

776. Horns. Macartney collection.

777. Horns.   do.

Antilope scripta. PALL. The Harnessed Antelope.

778. Head and horns, wanting the lower jaw.

Antilope Kevella. GMEL.

779. Skeleton. Presented by Mr Baker, Naturalist, Cambridge.

Antilope cervicapra. PALL. The Indian Antelope.

780. Head of a male. Presented by Prof. Clark.

781. Right horn. [Bell collection.] Presented by Prof. Clark.

Antilope picta. PALL. The Nylghau.

782. Head and horns. Macartney collection.

Antilope Oreas. PALL. The Eland.

Cervical vertebræ ...	7.	
Dorsal	do.	... 13.
Lumbar	do.	... 6.
Sacral	do.	... 5.
Caudal	do.	... 4 (the rest are wanting).

783. Skeleton. Presented by Prof. Clark.

784. Frontlet and horns. [Bell collection.]

Presented by Prof. Clark.

Antilope rupicapra. PALL. **The Chamois.**

785.   Frontlet and horns.   Presented by Prof. Clark.

Antilope Lervia. PALL.

786.   Head and horns, with the natural integuments.
                        Presented by Huddlestone Stokes, Esq.

Genus *Capra.* LINNÆUS.

Dental formula the same as the last.

Capra Hircus. L. **The Goat.**

> Cervical vertebræ ... 7.
> Dorsal   do.   ... 13.
> Lumbar   do.   ... 6.
> Sacral   do.   ... 4.
> Caudal   do.   ... 11.

787.   Skeleton.   Presented by Prof. Clark.
788.   Head of a variety from Ceram I.   [French Exp.]
                        Presented by Prof. Clark.
789.   Head and horns of a variety from Ceram I.   [French Exp.]
                        Presented by Prof. Clark.
790.   Head, with the right horn, of a variety from Zamboangan,
Mindanao I.   [French Exp.]   Presented by Prof. Clark.
791.   Head of a variety from Valparaiso in Chili.   [French Exp.]
                        Presented by Prof. Clark.

Capra Ibex. L. **The Steinbok, or Bouquetin.**

792.   Head, with a section through the core of the left horn, to show
the connection of the frontal sinus with the cavity of its interior.
                        Presented by Prof. Clark.
793.   Head and horns.   Bought at Chamounix in 1815.
                        Presented by Prof. Clark.
794.   Head and horns.   Macartney collection.
795.   Left horn.
796.   Right horn.

*

70 MAMMALS.

Genus *Ovis*. LINNÆUS.

Ovis Aries. L. **The Sheep.**

[A] The Four-horned variety.

Cervical vertebræ ... 7.
Dorsal   do.   ... 13.
Lumbar   do.   ... 6.
Sacral   do.   ... 5.
Caudal   do.   ... 12.

797. Skeleton. [Bell collection.] Presented by Prof. Clark.

798. Head. The fourth horn on the left side indicates a subdivision. [Bell collection.] Presented by Prof. Clark.

799. Head and horns, wanting the lower jaw and the premaxillary bones. [Bell collection.] Presented by Prof. Clark.

800. Head and horns. From Zamboangan, Mindanao I. [French Exp.] Presented by Prof. Clark.

[B] The Domestic variety.

801. Head, articulated according to the vertebral system of Prof. Owen. Presented by Prof. Clark.

802. Head. [Bell collection.] Presented by Prof. Clark.

803. Head and horns. Presented by Prof. Clark.

804. Head. [Bell collection.] Presented by Prof. Clark.

805. Horns of a male. Presented by J. W. Clark, M.A. Trin.

806. Bones complete. Presented by Prof. Clark.

807. Head and horns, wanting the premaxillary bones. From Timor I. [French Exp.] Presented by Prof. Clark.

808. Head. From Timor I. [French. Exp.]
Presented by Prof. Clark.

Genus *Bos*. LINNÆUS.

Dental formula the same as the last.

Bos Taurus. L. **The Ox.**

809. Head and horns. [Bell collection.] Presented by Prof. Clark.

810. Horns of a bull.
Presented by Rev. J. S. Henslow, M.A. St John's College.

811. First and second vertebra. Presented by Prof. Clark.

812. Skeleton of a monstrous calf: showing the coalition of the two bodies by the second cervical vertebra in each forming a single mass.
Presented by Prof. Clark.

Bos Indicus. L. **The Zebu.**

Cervical vertebræ ... 7.
Dorsal   do.   ... 13.
Lumbar   do.   ... 6.
Sacral   do.   ... 4.
Caudal   do.   ... 17 (the rest are wanting).

813. Skeleton. From the Zoological Society's gardens.
Presented by Prof. Clark.

814. Young head, from Wentworth Park.
Presented by Earl Fitzwilliam.

815. Head, from the same locality. Presented by Earl Fitzwilliam.

Bos bubalis. L. **The Buffalo.**

816. Head and horns. [Bell collection.] Presented by Prof. Clark.

Bos Caffer. SPARMANN. **The Cape Buffalo.**

Cervical vertebræ ... 7.
Dorsal   do.   ... 13.
Lumbar   do.   ... 6.
Sacral   do.   ... 4.
Caudal   do.   ... 17 (the rest are wanting).

817. Skeleton. Purchased by subscription.

818. Head and horns, with the natural integuments. [Bell collection.] Presented by Prof. Clark.

Bos Arni. BLUMENBACH. **The Indian Buffalo.**

819. Cranium and horns. Presented by Huddlestone Stokes, Esq.

820. Another specimen, with its horns and a larger portion of the cranium. Presented by Huddlestone Stokes, Esq.

821. Frontlet and horns. do.

822. Frontlet and horns. do.

Bos Primigenius. BOJAN. (*Fossil.*)

823. Cranium and horns.
Presented by Rev. J. S. Henslow, M.A. St John's College.

824. Right radius and ulna. From the same locality in Ireland as the *Cervus megaceros.*

Bos Longifrons. OWEN. (*Fossil.*)

825. Right half of the cranium, with a portion of its horn. From the same locality in Ireland as the *Cervus megaceros.*

826. A right radius and ulna.

827. The cuboid, navicular, and external cuneiform bones, anchylosed.

Fossil bones from the gravel near Cambridge:

828. Inferior portion of the right scapula.

829. Inferior portion of the right humerus.

830. A sacral vertebra.

831. Fragment of the extremity of the right radius.
Presented by Prof. Clark.

Order VI. EDENTATA.

Family XX. EFFODIENTIA.

Genus *Orycteropus.* GEOFFROY ST HILAIRE.

Orycteropus capensis. GEOFFR. The Cape Ant-eater.

Cervical vertebræ ...	7.
Dorsal do. ...	12.
Lumbar do. ...	8.
Sacral do. ...	6.
Caudal do. ...	26.

832. Skeleton. Purchased by subscription.

Genus *Dasypus.* L.

Dasypus sexcinctus. L. **The Six-banded Armadillo.**

Cervical vertebræ ... 7.
Dorsal do. ... 11.
Lumbar do. ... 3.
Sacral do. ... 9.
Caudal do. ... 13.

833. Skeleton. Presented by Prof. Clark.

Genus *Glyptodon.* OWEN. (*Fossil*).

Glyptodon clavipes. OWEN.

834. A portion of the carapace; part of the original specimen brought from South America by Sir W. Parish.
Presented by the Royal College of Surgeons of England.

Genus *Mylodon.* OWEN. (*Fossil*).

Mylodon Darwinii. OWEN.

835. A cast of the right ramus of the lower jaw: figured and described in Owen's "Zoology of the Voyage of the Beagle," p. 63.
Presented by the Royal College of Surgeons of England.

Genus *Scelidotherium.* OWEN. (*Fossil*).

Scelidotherium leptocephalum. OWEN.

Casts of the following bones: described in Owen's "Zoology of the Voyage of the Beagle," p. 73 sq.
Presented by the Royal College of Surgeons of England.

836. The skull and right stylohyal bone.

837. Vertebræ and fragments of ribs.

838. Four vertebræ.

839. The sacrum.

840. The left scapula.

10

841.   Proximal portion of the left humerus, with fragments of scapula.

842.   The right femur.

843.   The left femur.

Family XXI. TARDIGRADA.

Genus Bradypus. LINNÆUS.

Bradypus tridactylus.   L.   **The Three-fingered Sloth.**

     Cervical vertebræ ...   9.
     Dorsal   do.   ... 15.
     Lumbar   do.   ...   4.
     Sacral   do.   ...   6.
     Caudal   do.   ...   5 (the rest are wanting).

844.   Skeleton.   [Bell collection]   Presented by Prof. Clark.

Order VII. RODENTIA.

Family XXII. DUPLICIDENTATA.

Genus Lepus. L.

Dental formula: $i. \frac{2-2}{1-1}$, $p. \frac{3-3}{2-2}$, $m. \frac{3-3}{3-3} = 28.$

Lepus timidus.   L.   **The Hare.**

     Cervical vertebræ ...   7.
     Dorsal   do.   ... 12.
     Lumbar   do.   ...   7.
     Sacral   do.   ...   3.
     Caudal   do.   ...   7 (the rest are wanting).

845.   Skeleton.   [Bell collection.]   Presented by Prof. Clark.

846.   Head.

847.   Head reversed to show the inferior surface and teeth.   } Presented by Prof. Clark.

Lepus hibernicus.   BELL.   **The Irish Hare.**

848.   Skeleton.   Brookes collection.

849.   Skeleton.   Presented by Prof. Clark.

Lepus cuniculus. L. **The Rabbit.**

850. Head.
851. Head reversed to show the inferior surface and teeth. } Presented by Prof. Clark.

Family XXIII. SUBUNGULATA.

Genus *Cavia.* ILLIGER.

Dental formula: $i.\ \frac{1-1}{1-1},\ p.\ \frac{1-1}{1-1},\ m.\ \frac{3-3}{3-3}=20.$

Cavia cobaya. SCHREB. **The Guinea-Pig.**

Cervical vertebræ ... 7.
Dorsal do. ... 13.
Lumbar do. ... 6.
Sacral do. ... 4.
Caudal do. ... 6.

852. Skeleton.
Presented by H. J. H. Bond, M.D. Corpus Christi College.
853. Skeleton. Macartney collection.
854. Head.
855. Head reversed to show the inferior surface and teeth. } Presented by Prof. Clark.

Genus *Hydrochœrus.* BRISSON.

Hydrochœrus capybara. DESM. **The Capybara.**

856. Head, wanting the lower jaw. Presented by Prof. Clark.

Genus *Cœlogenys.* CUVIER.

Cœlogenys subnigra. CUV. **The Brown Paca.**

Cervical vertebræ ... 7.
Dorsal do. ... 12.
Lumbar do. ... 6.
Sacral do. ... 1 (others would be anchylosed shortly).
Caudal do. ... 9.

857. Skeleton of a very young animal. Brookes collection.

Family XXIV. ACULEATA.

Genus *Hystrix*. LINNÆUS.

Dental formula the same as the last.

Hystrix javanica. CUV. **The Porcupine of Java.**

Cervical vertebræ ... 7.
Dorsal   do.   ... 15.
Lumbar   do.   ... 4.
Sacral   do.   ... 4.
Caudal   do.   ... 14.

858. Skeleton. [Bell collection.] Presented by Prof. Clark.

Hystrix cristata. L. **The Crested** or **Common Porcupine.**

Cervical vertebræ ... 7.
Dorsal   do.   ... 14.
Lumbar   do.   ... 5.
Sacral   do.   ... 4.
Caudal   do.   ... 7.

859. Skeleton. The animal died in Wombwell's Menagerie.
Presented by Prof. Clark.

Hystrix hirsutirostris. BRANDT. **The Asiatic Porcupine.**

860. Head. [Bell collection.] Presented by Prof. Clark.

Hystrix (species uncertain).

861. Old head: possibly of an exotic variety of *H. cristata*.

Family XXV. PALMIPEDIA.

Genus *Castor*. LINNÆUS.

Castor fiber. L. **The Beaver.**

Dental formula the same as the last.

862. Head, wanting a part of the occiput. Harwood collection.
863. Head, wanting the right moiety of the lower jaw. [Bell collection.] Presented by Prof. Clark.

Genus *Myopotamus*. COMMERSON.

Dental formula the same as the last.

Myopotamus coypus. CUVIER. **The Coypu.**

Cervical vertebræ ... 7.
Dorsal do. ... 12.
Lumbar do. ... 6.
Sacral do. ... 4.
Caudal do. ... 14 (the rest of the normal 21 are wanting).

864. Skeleton, from the Zoological Society's Gardens.

Presented by Prof. Clark.

Family XXVI. MURINA.

Genus *Mus*. LINNÆUS.

Dental formula: $i. \frac{1-1}{1-1}, \quad p. \frac{1-1}{1-1}, \quad m. \frac{2-2}{2-2} = 16.$

Mus decumanus. L. **The Brown, or Norway, Rat.**

Cervical vertebræ ... 7.
Dorsal do. ... 13.
Lumbar do. ... 6.
Sacral do. ... 2.
Caudal do. ... 29.

865. Skeleton.

Presented by H. J. H. Bond, M.D.

866. Skeleton. There are three sacral vertebræ. [Bell collection.]

Presented by Prof. Clark.

867. Head, disarticulated. Presented by H. J. H. Bond, M.D.

868. Head.

869. Head reversed, to show the inferior surface and teeth.

Presented by Prof. Clark.

Mus musculus. L. **The Common Mouse.**

870. Group of three skeletons in different attitudes.

Presented by Prof. Clark.

871. Skeleton. [Bell collection.] Presented by Prof. Clark.

872. A dried specimen.

Genus *Arvicola*. Lacepede.

Dental formula the same as the last.

Arvicola amphibius. Desmar. **The Water Vole.**

873. Head.

874. Head reversed, to show the inferior surface and teeth.}

       Presented by Prof. Clark.

Arvicola agrestis. Fleming. **The Meadow Mouse.**

875. Head. Presented by Prof. Clark.

Family XXXI. Sciurina.

Genus *Arctomys*. Schreber.

Dental formula: $i. \frac{1-1}{1-1}, \; p. \frac{1-1}{1-1}, \; m. \frac{4-4}{3-3} = 22.$

Arctomys marmota. Schreb. **The Marmot.**

    Cervical vertebræ ... 7.
    Dorsal do. ... 12.
    Lumbar do. ... 7.
    Sacral do. ... 4.
    Caudal do. ... 19.

876. Skeleton. Presented by Prof. Clark.

Genus *Pteromys*. Cuvier. **Flying Squirrels.**

Dental formula the same as the last.

Pteromys volucella. Gmel.

877. Head. Brookes collection.

Genus *Sciurus*. L.

Dental formula as in *Pteromys*.

Sciurus bicolor. Sparmann. **The Javan Squirrel.**

878. A stuffed specimen. Presented by T. Horsfield, M.D.

Sciurus caniceps. GRAY. **The Ashy-headed Squirrel.**

879. Head. Presented by T. Horsfield, M.D.

Sciurus Carolinensis. GMEL. **The Carolina Squirrel.**

880. Head. Brookes collection.
881. Head reversed, to show the inferior surface and teeth.

Presented by Prof. Clark.

Sciurus vulgaris. L. **The Common Squirrel.**

Cervical vertebræ ... 7.
Dorsal do. ... 12.
Lumbar do. ... 7.
Sacral do. ... 3.
Caudal do. ... 23.

882. Skeleton.
883. Skeleton.
884. Head.   } Presented by
885. Head reversed to show the inferior surface } Prof. Clark.
     and teeth.

Genus *Tamias.* ILLIGER. **Ground Squirrels.**

Dental formula: $i. \frac{1-1}{1-1}$, $p. \frac{1-1}{1-1}$, $m. \frac{3-3}{3-3} = 20.$

Tamias striatus. ILLIG.

886. Head. Brookes collection.

Order VIII. FERÆ.

Family XXXII. PINNIPEDIA.

Genus *Trichechus*. LINNÆUS.

Dental formula: $i. \frac{3-3}{2-2},\ \ c. \frac{1-1}{0-0},\ \ m. \frac{4-4}{4-4} = 28.$

Trichechus Rosmarus. L. **The Morse, or Walrus.**

887.	Head.	
888.	Older head.	
889.	Right half of the cranium.	
890.	Left half of the cranium.	[Bell collection.]
891.	The lower jaw.	Presented by Prof. Clark.
892.	Muzzle, with enormous tusks.	
893.	Os penis.	

Genus *Cystophora*. NILSSON.

Dental formula: $i. \frac{2-2}{1-1},\ \ c. \frac{1-1}{1-1},\ \ m. \frac{5-5}{5-5} = 30.$

Cystophora proboscidea. NILLS. **The Sea Elephant.**

894.   A dorsal vertebra.   Presented by Prof. Owen.

Cystophora cristata. CUV. **The Hooded Seal.**

895.   Head. [Bell collection.]   Presented by Prof. Clark.
896.   Younger head. [Bell collection.]   Presented by Prof. Clark.

Genus *Stenorhynchus*. CUV.

Dental formula: $i. \frac{2-2}{2-2},\ \ c. \frac{1-1}{1-1},\ \ m. \frac{5-5}{5-5} = 32.$

Stenorhynchus serridens. OWEN. **The Crab-eating Seal.**

897.   Adult head.   Antarctic Seas.   [French Exp.]
                                Presented by Prof. Clark.

Stenorhynchus leptonyx. BLAINVILLE. **The Sea Leopard.**

898. Adult head. Antarctic Seas. [French Exp.]
Presented by Prof. Clark.

899. Left half of lower jaw, with the teeth complete. [Bell collection.] Presented by Prof. Clark.

Genus *Halichœrus.* NILSSON.

Dental formula: $i.\ \dfrac{3-3}{2-2},\ c.\ \dfrac{1-1}{1-1},\ m.\ \dfrac{5-5}{5-5}=34.$

Halichœrus gryphus. NILSS. **The Grey Seal.**

900. Skeleton of an old individual, caught in the fishing nets off Colchester, being blind.

Cervical vertebræ ...	7.
Dorsal do. ...	16.
Lumbar do. ...	4.
Sacral do. ...	4.
Caudal do. ...	13.

Presented by Prof. Clark.

901. Young skeleton. In this specimen there are fifteen dorsal vertebræ, five lumbar, and seventeen between the last lumbar and the end of the tail, which seems perfect. It is worthy of remark that both these specimens have six molars on the right side of the upper jaw. Macartney collection.

Phoca vitulina. L. **The Common Seal.**

902. Head, wanting the malar bones, and the lower jaw. [Bell collection.] Presented by Prof. Clark.

903. Head of a young specimen, shewing the interior of the auditory bulla on the left side. On the right side the temporal bone has been removed, and also the malar bone. The dentition is complete. Presented by Prof. Clark.

11

Genus *Arctocephalus*. F. CUVIER.

Arctocephalus ursinus. F. CUV.

904.   Head, from the Falkland Islands. [French Exp.]

Presented by Prof. Clark.

905.   Left half of lower jaw, with the teeth complete. [Bell collection.] Presented by Prof. Clark.

906.   Left half of upper jaw of the same specimen. [Bell collection.]

Presented by Prof. Clark.

Genus *Otaria*. PERON.

Dental formula : $i. \dfrac{3-3}{2-2}, \quad c. \dfrac{1-1}{1-1}, \quad m. \dfrac{6-6}{5-5} = 36.$

Otaria leonina. PER. **Southern Sea Bear,** Byron's **Sea Lion.**

907.   Head of a male. A transverse section had been made to show the cavity of the cranium. The animal has been killed by a blow on the nose, which has fractured the bones between the orbits. From the coast of Chili. [French Exp.]

Presented by Prof. Clark.

908.   Head. This specimen also has the cavity of the cranium shewn by a transverse section. The animal was killed by a fracture of the nose, which however has exposed the nasal passages on the right side only. From the Straits of Magellan. [French Exp.] Presented by Prof. Clark.

909.   The mutilated head of an old specimen. It wants the lower jaw, and is much weather-worn. The sutures are obliterated. From the coast of Araucaria. [French Exp.]

Presented by Prof. Clark.

Phocidæ (Species uncertain).

910.   A right radius, from the Straits of Magellan. [French Exp.]

Presented by Prof. Clark.

911. Head of a young specimen. The right side of the lower jaw is wanting.

Dental formula: $i. \frac{3-3}{2-2}$, $c. \frac{1-1}{1-1}$, $m. \frac{6-6}{5-5}$.

This skull has a marked development of the cranial cavity, also a marked post-orbital process on the frontal bones as well as an anterior on the superior maxillary bones. The two halves of the frontal bone unite to form an entering angle between the nasals, which is contrary to what is seen in the other specimens in this Museum, in which the united nasal bones form an acute angle which is received between the produced parts of the frontal bone. The inferior orbital process is formed entirely on the malar bone. The palate bones terminate behind by a round edge. This remarkable skull suits *Cystophora* more nearly than any other known genus, except in the number of its molar teeth and the great width of its cranial cavity. Its dental formula is the same as in *Otaria*. (Bell collection.) Presented by Prof. Clark.

Family XXXIII. FELINA.

Genus *Felis*. L.

Dental formula: $i. \frac{3-3}{3-3}$, $c. \frac{1-1}{1-1}$, $p. \frac{3-3}{2-2}$, $m. \frac{1-1}{1-1} = 30$.

Felis Leo. L. **The Lion.**

Cervical vertebræ ...	7.
Dorsal	do.	... 13.
Lumbar	do.	... 7.
Sacral	do.	... 3.
Caudal	do.	... 25.

912. Skeleton of a male African lion. The bones of the extremities are thickened by osseous deposits from periosteal inflammation. Brookes collection.

913. Head of a male African lion. Presented by Prof. Clark.

914. Head of a male African lion. He was shot at Astley's Theatre on account of his ferocity.
Presented by Mr Baker, Naturalist, Cambridge.

915. Head of an African lioness. Presented by Prof. Clark.

916. Head of an Asiatic lion.   Presented by Prof. Clark.

917. Head of an Asiatic lion.   do.

918. Bones of the lion, No. 914.
                    Presented by Mr Baker, Naturalist, Cambridge.

919. Skeleton of a hybrid between a lion and a tigress, bred in Wombwell's menagerie.   Presented by G. Kemp, M.D.

Felis Tigris. L. **The Tiger.**

Cervical vertebræ... 7.
Dorsal   do.   ... 13.
Lumbar   do.   ... 7.
Sacral   do.   ... 3.
Caudal   do.   ... 25.

920. Skeleton of a Royal tiger of Bengal, from the Zoological Society's Gardens.   Presented by Prof. Clark.

921. Head.   Presented by the Master and Fellows of Trinity College.

922. Head.   Presented by Prof. Clark.

923. Head.   [Bell collection].   Presented by Prof. Clark.

924. Head.   Presented by the Marquis Spineto.

925. Head.   Presented by Huddlestone Stokes, Esq.

926. Head.   do.

927. Left femur.   Brookes collection.

928. Left hind leg.   Harwood collection.

929. Lower jaw.   [Bell collection.]   Presented by Prof. Clark.

Felis Leopardus. TEMMINCK. **The Leopard.**

930. Head.

931. Head.   [Bell collection.]

932. Head, with the atlas attached.   Presented by Prof. Clark.

933. Young head.

Felis pardalis. LAURENT. **The Ocelot.**

934. Head.   [Bell collection.]   Presented by Prof. Clark.

935. Skin of the head, dried, with the incisor and canine teeth. [Bell collection.]   Presented by Prof. Clark.

Felis concolor. L. **The Red Puma.**

Cervical vertebræ... 7.
Dorsal   do.   ... 13.
Lumbar   do.   ... 7.
Sacral   do.   ... 3.
Caudal   do.   ... 18 (the rest are wanting).

936. Skeleton. From the Zoological Society's Gardens.

Presented by Prof. Clark.

Felis Javanensis. DESM. **The Kuwuk.**

937. Head. Presented by Prof. Clark.

Felis catus. L. **The Domestic Cat.**

938. Skeleton. Presented by H. J. H. Bond, M.D.

939. Skeleton of a tailless variety. From the Zoological Society's gardens. Presented by Prof. Clark.

940. Dried specimen, found in a house in Cambridge when the wainscotting was taken down.

Presented by Mr T. Boning, Cambridge.

941. Head. ⎫
942. Head wanting the lower jaw. ⎬ Presented by Prof. Clark.
943. Head and bones of an old male. ⎭

944. Head, with the first two cervical vertebræ attached. [Bell collection.] Presented by Prof. Clark.

945. Head, disarticulated. Purchased by the University.

Family XXXIV. VIVERRINA.

Genus *Hyæna.* BRISSON.

Dental formula: $i. \frac{3-3}{3-3}, \quad c. \frac{1-1}{1-1}, \quad p. \frac{4-4}{3-3}, \quad m. \frac{1-1}{1-1} = 34.$

Hyæna striata. ZIMMERMAN. **The Hyæna.**

Cervical vertebræ... 7.
Dorsal   do.   ... 15.
Lumbar   do.   ... 5.
Sacral   do.   ... 2.
Caudal   do.   ... 12 (the rest are wanting).

946.   Skeleton.   Presented by Prof. Clark.

947.   Very old head.   Macartney collection.

Hyæna Spelæa.   GOLDFUSS.   Cave Hyæna.   (Fossil.)

Teeth and fragments of bones from Kent's cavern.*

Macartney collection.

948.   Fragment of the left maxillary bone, with the 3rd and 4th molar.

949.   Fragment of the left ramus of the lower jaw, with the 1st, 2nd, and 3rd molars.

950.   A canine Tooth.

951.   A canine Tooth.

952.   A canine Tooth.

953.   First upper molar, right side, with a fragment of the maxillary bone.

954.   Fourth upper molar, right side.

955.   Fourth upper molar, left side.

956.   Third upper molar, left side.

957.   Second lower molar, right side.

958.   Third lower molar, right side.

959.   Fourth lower molar, left side.

960.   Fourth lower molar, right side.

961.   Fragment of the long bone of a Mammal, gnawed by the hyænas in the above cavern.

962.   A similar fragment.

Genus Viverra.   LINNÆUS.

Dental formula : $i. \frac{3-3}{3-3}, \quad c. \frac{1-1}{1-1}, \quad p. \frac{4-4}{4-4}, \quad m. \frac{2-2}{2-2} = 40.$

Viverra Genetta.   L.   The Ring-tailed Civet Cat.

Cervical vertebræ ...   7.
Dorsal   do.   ... 13.
Lumbar   do.   ...   7.
Sacral   do.   ...   3.
Caudal   do.   ... 24 (the rest are wanting).

963.   Skeleton.   [Bell collection.]   Presented by Prof. Clark.

---

* For an account of Kent's cavern see "Cavern Researches," edited from the MSS. of the Rev. J. MacEnery, one of its first explorers, by E. Vivian, Esq. Lond. 1859.

Viverra malaccensis. GMEL. **The Rasse.**

964. Head. [Bell collection.] Presented by Prof. Clark.

Genus *Herpestes.* ILLIGER.

Dental formula the same as the last.

Herpestes Ichneumon. ILLIG. **The Ichneumon.**

Cervical vertebræ... 7.
Dorsal   do.   ... 14.
Lumbar do.   ... 6.
Sacral   do.   ... 3.
Caudal   do.   ... 26 (the rest are wanting.)

965. Skeleton. [Bell collection.] Presented by Prof. Clark.

Family XXXV. CANINA.

Genus *Canis.* LINNÆUS.

Dental formula: $i. \frac{3-3}{3-3},\ c. \frac{1-1}{1-1},\ p. \frac{4-4}{4-4},\ m. \frac{2-2}{3-3} = 42.$

Canis Vulpes. L. **The Fox.**

Cervical vertebræ... 7.
Dorsal   do.   ... 13.
Lumbar do.   ... 7.
Sacral   do.   ... 3.
Caudal   do.   ... 16 (the rest are wanting).

966. Skeleton. Presented by G. M. Humphrey, M.D.

967. Head. Presented by Prof. Clark.

968. Head, showing the cavity of the cranium. [Bell collection.]
Presented by Prof. Clark.

969. Head.⎫ Presented by Prof. Clark.
970. Head.⎭

971. Cranium of a young specimen. Presented by Prof. Clark.

972. Cranium of a somewhat older individual.
Presented by Prof. Clark.

973. Portion of the head of the Arctic variety, showing the teeth.
[Bell collection.] Presented by Prof. Clark.

Canis Lupus. L. **The Wolf.**

974.   Skeleton.   From the Zoological Society's Gardens.
$\qquad$ Presented by Prof. Clark.

975.   Head, the zygomatic arches are broken.   [Bell collection.]
$\qquad$ Presented by Prof. Clark.

976.   Head, dried.   [Bell collection.]   Presented by Prof. Clark.

Canis familiaris. L. **The Dog.**

977.   Skeleton of "Miller," a Newfoundland Dog.   Macartney collection.

978.   Bones of an Isle of Skye Terrier.   Presented by Prof. Clark.

979.   Head of "Muschean," a Skye Terrier, belonging to Prof. Clark.   Presented by Prof. Clark.

980.   Head of a Skye Terrier.⎫
981.   Head of a Skye Terrier.⎬   Presented by Prof. Clark.
982.   Head of a Skye Terrier.⎭

983.   Head of "Toozie," a Skye Terrier bitch, belonging to Prof. Clark.   Presented by Prof. Clark.

984.   The skull of a Skye Terrier shortly after birth.
985.   do.   do.   a little younger.
986.   do.   do.   still younger.
$\qquad$ Presented by Prof. Clark.

987.   Bones of a Terrier.   Presented by Prof. Clark.

988.   Bones of a hybrid between a Terrier and another.
$\qquad$ Presented by Miss Bond.

989.   Head of a Greyhound.   Presented by Prof. Clark.

990.   Head of a Spaniel.   do.

991.   Head of a young Pointer, disarticulated.
$\qquad$ Presented by Prof. Clark.

992.   Head of a large Mongrel, with the sutures obliterated.
$\qquad$ Presented by Prof. Clark.

993.   Head.   [Bell collection.]   Presented by Prof. Clark.

994.   Head.   do.   do.

995.   Head.   Presented by Prof. Clark.

996.   Head.   do.

997. Head. Presented by Prof. Clark.

998. Head, wanting the nasal bones. do.

999. Head, wanting the lower jaw, of a variety from the Bay of Islands, New Zealand. [French Exp.] Presented by Prof. Clark.

1000. The charred skull of a Dog. He was left in guard of a house at Pâh d'Acaroa, New Zealand, and when it was fired refusing to leave it, was consumed together with it.

[French Exp.] Presented by Prof. Clark.

1001. Head, with the occipital and parietal bones removed to display the cavity of the cranium, and the foramina of the ethmoid bone. Presented by Prof. Clark.

Head, divided by two sections into three portions.

1002. (A) Shows the posterior part of the osseous tentorium, and the foramen magnum.

1003. (B) Through the auditory bulla on each side, exposing the osseous organ of hearing, and the anterior part of the bony tentorium.

1004. (C) Shows the cavity for the anterior lobes of the brain, the optic and the olfactory foramina.

Presented by Prof. Clark.

Head divided by two sections into three portions.

1005. (A) Shows the osseous tentorium, the section of the semicircular canals of the ear, and the posterior portion of the bulla.

1006. (B) The anterior part of the bulla, and middle regions of the cranial cavity, together with the frontal and sphenoidal cells.

1007. (C) The anterior cerebral cavity with the ethmoidal lamina and frontal cells.

1008. Head, divided in the mid plane, to show the cavity of the cranium. In one of the halves the septum narium is seen, in the other the convolutions of the turbinated bones.

Presented by Prof. Clark.

1009. Head, with the component bones disarticulated, and united by wires. Presented by Prof. Clark.

Family XXXVI. MUSTELINA.

Genus *Lutra*. RAY.

Dental formula: $i. \frac{3-3}{3-3}, \quad c. \frac{1-1}{1-1}, \quad p. \frac{4-4}{3-3}, \quad m. \frac{1-1}{2-2} = 36.$

Lutra vulgaris. ERXLEB. **The Common Otter.**

Cervical vertebræ ... 7.
Dorsal   do.   ... 15.
Lumbar   do.   ... 6.
Sacral   do.   ... 3.
Caudal   do.   ... 25.

1010.   Skeleton. Macartney collection.

1011.   Bones of an animal killed in Bedfordshire.}   Presented by
1012.   Do. of another. }   W. Drake, Esq.

1013.   Old head. Presented by Prof. Clark.

Lutra (*Species uncertain*).

1014.   Left hind foot, dried. Harwood collection.

Genus *Aonyx*. LESSON.

Aonyx leptonyx. LESSON.

1015.   Head. [Bell collection.] }   Presented
1016.   Head reversed, to show the teeth.   do.} by Prof. Clark.

1017.   A stuffed specimen. Presented by Thomas Horsfield, M.D.

Genus *Mustela*. LINNÆUS.

Dental formula: $i. \frac{3-3}{3-3}, \quad c. \frac{1-1}{1-1}, \quad p. \frac{3-3}{3-3}, \quad m. \frac{1-1}{2-2} = 34.$

Mustela putorius. L. **The Polecat, Foumart, or Fitchet Weasel.**

Cervical vertebræ ... 7.
Dorsal   do.   ... 14.
Lumbar   do.   ... 6.
Sacral   do.   ... 3.
Caudal   do.   ... 17.

1018.   Skeleton. Presented by H. J. H. Bond, M.D.

1019. Skeleton. Presented by G. M. Humphry, M.D.

1020. Skeleton. [Bell collection.] Presented by Prof. Clark.

1021. Old skull. }
1022. Younger skull. } [Bell collection.] Presented by Prof. Clark.

Mustela erminea. L. **Stoat, or Greater Weasel.**

1023. Skeleton. Presented by Prof. Clark.

Mustela vulgaris. L. **The Common Weasel.**

1024. Skeleton. Presented by H. J. H. Bond, M.D.

1025. Head of a male. }
1026. Head of a female. } [Bell collection.] Presented by Prof. Clark.

Mustela furo. L. **The Ferret Weasel.**

1027. Skeleton. Macartney collection.

Genus *Martes.* RAY.

Dental formula : $i. \frac{3-3}{3-3}, \quad c. \frac{1-1}{1-1}, \quad p. \frac{4-4}{4-4}, \quad m. \frac{1-1}{2-2} = 38.$

Martes foina. GMELIN. **The Common Marten.**

1028. Head. [Bell collection.] Presented by Prof. Clark.

Martes abietum. RAY. **The Pine Marten.**

1029. Head, wanting the lower jaw. [Bell collection.]
Presented by Prof. Clark.

Genus *Galictis.* BELL.

Dental formula : $i. \frac{3-3}{3-3}, \quad c. \frac{1-1}{1-1}, \quad p. \frac{3-3}{3-3}, \quad m. \frac{1-1}{2-2} = 34.$

Galictis vittata. BELL.

Cervical vertebræ ... 7.
Dorsal   do.   ... 15.
Lumbar   do.   ... 5.
Sacral   do.   ... 3.
Caudal   do.   ... 18.

1030.   Skeleton.   The animal was kept by Mr Bell as a pet, and has been described by him in the *Transactions of the Zoological Society* for 1839, pp 201—208.   [Bell collection.]

Presented by Prof. Clark.

Genus *Mephitis*. CUVIER.

Dental formula the same as the last.

Mephitis mustelina.   CUV.   **The Striped Weasel of Africa.**

1031.   The forepart of the head, taken from a stuffed specimen. [Bell collection.]   Presented by Prof. Clark.

Genus *Mydaus*. F. CUVIER.

Dental formula the same as the last.

Mydaus meliceps.   F. CUV.

1032.   A stuffed specimen.   Presented by Thos. Horsfield, M.D.

Genus *Meles*. CUVIER.

Dental formula: $i. \dfrac{3-3}{3-3}, \quad c. \dfrac{1-1}{1-1}, \quad p. \dfrac{4-4}{4-4}, \quad m. \dfrac{1-1}{2-2} = 38.$

Meles taxus.   FLEMING.   **The Badger.**

Cervical vertebræ ...   7.
Dorsal   do.   ... 15.
Lumbar   do.   ... 5.
Sacral   do.   ... 2.
Caudal   do.   ... 16 (a few are wanting).

1033.   Skeleton.   [Bell collection].   Presented by Prof. Clark.
1034.   Skeleton.   Presented by Prof. Clark.
1035.   Old head.   Macartney collection.
1036.   Adult head.   [Bell collection.]   Presented by Prof. Clark.
1037.   Younger head.⎱
1038.   Young head. ⎰ Presented by Prof. Clark.

Genus *Mellivora*. STORR.

Dental formula: $i.\ \frac{3-3}{3-3},\ c.\ \frac{1-1}{1-1},\ p.\ \frac{3-3}{3-3},\ m.\ \frac{1-1}{1-1}=32.$

Mellivora capensis. GMELIN. **The Ratel.**

Cervical vertebræ ... 7.
Dorsal   do.   ... 15.
Lumbar   do.   ... 4.
Sacral   do.   ... 3.
Caudal   do.   ... 17.

1039.   Skeleton. [Bell collection]. Presented by Prof. Clark.

Family XXXVII. URSINA.

Genus *Ursus*. LINNÆUS.

Dental formula: $i.\ \frac{3-3}{3-3},\ c.\ \frac{1-1}{1-1},\ p.\ \frac{4-4}{4-4},\ m.\ \frac{2-2}{3-3}=42.$

Ursus Arctos. L. **The Brown Bear of Europe.**

Cervical vertebræ ... 7.
Dorsal   do.   ... 14.
Lumbar   do.   ... 6.
Sacral   do.   ... 6.
Caudal   do.   ... 7.

1040.   Skeleton. Purchased from a menagerie.

Presented by Prof. Clark.

Ursus Americanus. PALLAS. **The Black Bear of America.**

Cervical vertebræ ... 7.
Dorsal   do.   ... 15.
Lumbar   do.   ... 5.
Sacral   do.   ... 4.
Caudal   do.   ... 9.

1041.   Young skeleton, which may account for the anomalous character of the vertebral formula. The British Museum Catalogue gives fourteen dorsal vertebræ, six lumbar, three sacral. From the Museum of J. P. Delafons, Esq. Presented by Prof. Clark.

Ursus maritimus. L. **The Polar Bear.**

1042. Head of a male, of unusual size. [Bell collection.]
Presented by Prof. Clark.

1043. Head of an adult male. [Bell collection.]
Presented by Professor Clark.

1044. Head. [Bell collection.] Presented by Prof. Clark.

1045. Younger head. [Bell collection.] Presented by Prof. Clark.

1046. Head. do. do.

Sections of head.

1047. (A) Longitudinal section, showing the proportion of the cranium to the face.

The other half of the head is divided by two transverse sections into three portions :

1048. (B) A section in front of the petrous part of the temporal bone, showing the osseous tentorium and cavity for the cerebellum.

1049. (C) A section near the cribriform lamella of the ethmoid bone, showing the cavities for the middle and anterior lobes of the brain.

1050. (D) The anterior portion of the section, showing the nasal passage and lamellæ of the turbinated bones. This also shows the dentition.

1051. A section of the head in the mid-plane, showing the osseous tentorium, the sphenoidal cells, the upper ethmoidal cells, and the septum narium. Presented by Prof. Clark.

1052. The two halves of the lower jaw of the preceding specimen.
Presented by Prof. Clark.

Ursus labiatus. Blainville. **The Long-lipped Bear.**

1053. Head.⎱
1054. Head.⎰ Presented by Huddlestone Stokes, Esq.

Genus *Helarctos.* Horsfield.

Helarctos Tibetanus. Horsf. **The Thibet Bear.**

1055. Head. [Bell collection.] Presented by Prof. Clark.

Ursidæ (*Species uncertain.*)

1056. Right femur of a bear.⎫
1057. Left femur of a bear. ⎭ Brookes collection.

1058. A lower jaw. [Bell collection.]

Presented by Prof. Clark.

Ursus spelœus. BLUMENBACH. **Great Cave Bear.** (*Fossil.*)

1059. Portion of a canine tooth.⎫ From Kent's Cavern. Macart-
1060. A molar tooth. ⎭ ney collection.

Genus *Procyon.* STORR.

Dental formula: $i. \frac{3-3}{3-3}, \quad c. \frac{1-1}{1-1}, \quad p. \frac{4-4}{4-4}, \quad m. \frac{2-2}{2-2} = 40.$

Procyon lotor. DESMAR. **The Racoon.**

Cervical vertebræ ... 7.
Dorsal do. ... 14.
Lumbar do. ... 6.
Sacral do. ... 3.
Caudal do. ... 15 (a few are wanting).

1061. Skeleton. [Bell collection.] Presented by Prof. Clark.
1062. Head. do. do.

Genus *Nasua.* STORR.

Dental formula the same as the last.

Nasua narica. **The Coati.**

Cervical vertebræ ... 7.
Dorsal do. ... 14.
Lumbar do. ... 6.
Sacral do. ... 3.
Caudal do. ... 25.

1063. Skeleton. From the Zoological Society's Gardens.

Presented by Prof. Clark.

1064. Old head.⎫
1065. Old head.⎭ [Bell collection.] Presented by Prof. Clark.

1066. Old head. Presented by Prof. Clark.

Family XXXVIII. TALPINA.

Genus *Talpa*. LINNÆUS.

Dental formula : $i. \frac{3-3}{3-3}, \quad c. \frac{1-1}{1-1}, \quad p. \frac{4-4}{4-4}, \quad m. \frac{3-3}{3-3} = 44.$

Talpa europæa. L. **The Common Mole.**

Cervical vertebræ ... 7.
Dorsal   do.   ... 13.
Lumbar   do.   ... 6.
Sacral   do.   ... 5.
Caudal   do.   ... 11.

1067. Skeleton. Harwood collection.

1068. Skeleton. Presented by Prof. Clark.

1069. A preparation, showing the connexions and form of the arms, clavicles, and scapulæ. Presented by Prof. Clark.

1070. Head.

1071. Head reversed, to show the teeth. } Presented by Prof. Clark.

Family XXXIX. SORICINA.

Genus *Sorex*. LINNÆUS.

Dental formula : $i. \frac{2}{2}, \quad c. \frac{1-1}{0-0}, \quad p. \frac{3-3}{2-2}, \quad m. \frac{4-4}{3-3} = 30.$

Sorex fodiens. PALLAS. **The Water Shrew.**

Cervical vertebræ ... 7.
Dorsal   do.   ... 14.
Lumbar   do.   ... 7.
Sacral   do.   ... 4.
Caudal   do.   ... 14.

1072. Skeleton. Presented by Prof. Clark.

Sorex araneus. L. **The Common Shrew.**

1073. Head. Presented by Prof. Clark.

Family XL. ERINACEINA.

Genus *Erinaceus.* LINNÆUS.

Dental formula: $i. \frac{3-3}{3-3}$, $c. \frac{0-0}{0-0}$, $p. \frac{4-4}{2-2}$, $m. \frac{3-3}{3-3} = 36.$

Erinaceus europæus. L. **The Hedgehog.**

Cervical vertebræ	...	7.
Dorsal do.	...	15.
Lumbar do.	...	6.
Sacral do.	...	4.
Caudal do.	...	11.

1074. Skeleton. Macartney collection.

1075. Skeleton.

1076. Head.

1077. Head reversed, to show the inferior surface and teeth.

} Presented by Prof. Clark.

Order IX. CHIROPTERA.

Family XLI. NYCTERINA.

Genus *Vespertilio.* GEOFFROY ST HILAIRE.

Dental formula: $i. \frac{2-2}{3-3}$, $c. \frac{1-1}{1-1}$, $p. \frac{3-3}{3-3}$, $m. \frac{3-3}{3-3} = 38.$

Vespertilio noctula. SCHREBER. **The Great Bat,** or **Noctule.**

Cervical vertebræ	...	7.
Dorsal do.	...	12.
Lumbar do.	...	4.
Sacral do.	...	4.
Caudal do.	...	10.

1078. Skeleton. Presented by H. J. H. Bond, M.D.

13

Vespertilio pipistrellus. Geoffroy. **The Common Bat.**

1079.   Skeleton. Macartney collection.

1080.   Skeleton. Presented by Prof. Clark.

Family XLII. Pterotocyna.

Genus *Pteropus.* Temminck.

Dental formula: $i. \frac{2-2}{2-2}, \quad c. \frac{1-1}{1-1}, \quad p. \frac{1-1}{3-3}, \quad m. \frac{3-3}{3-3} = 32.$

Pteropus rostratus. Cuv. **The Dog-Bat of Java.**

1081.   A stuffed specimen. } Presented by Thomas Horsfield, M.D.
1082.   The wings and head. }

Order X. Ptenopleura.

Family XLIII. Galeopitheci.

Genus *Galeopithecus.* Audebert.

Dental formula: $i. \frac{2-2}{2-3}, \quad c. \frac{1-1}{1-1}, \quad p. \frac{2-2}{2-2}, \quad m. \frac{3-3}{3-3} = 34.$

Galeopithecus varius. Audeb.

1083.   A stuffed specimen. Presented by Thomas Horsfield, M.D.

Order XI. Quadrumana.

Family XLIV. Lemurina.

Dental formula: $i. \frac{2-2}{2-2}, \quad c. \frac{1-1}{1-1}, \quad p. \frac{3-3}{3-3}, \quad m. \frac{3-3}{3-3} = 36.$

The following are of uncertain species.

1084.   Skeleton. [Bell collection.] Presented by Prof. Clark.

1085.   Skeleton, imperfect. Macartney collection.

1086.   A specimen dried and injected; it shows the teeth well.
                                         Harwood collection.

1087.   An adult skull. [Bell collection.] Presented by Prof. Clark.

Family XLV. SIMIÆ.

Phalanx II. HESPEROPITHECI.

Genus *Callithrix*. GEOFFROY ST HILAIRE.

Dental formula: $i. \frac{2-2}{2-2}$, $c. \frac{1-1}{1-1}$, $p. \frac{3-3}{3-3}$, $m. \frac{3-3}{3-3} = 36$.

Callithrix sciureus. GEOFFR. **The Marmoset.**

Cervical vertebræ ... 7.
Dorsal   do.   ... 12.
Lumbar   do.   ... 7.
Sacral   do.   ... 3.
Caudal   do.   ... 21.

1088. Skeleton. [Bell collection.] Presented by Prof. Clark.
1089. Head with the natural integuments.
                                Presented by Prof. Clark.

Callithrix (species uncertain).

1090. Adult skull, with the dentition complete.

Genus *Cebus*. CUVIER.

Dental formula the same as the last.

Cebus capucinus. ERXL. **The Capuchin Monkey.**

Cervical vertebræ ... 7.
Dorsal   do.   ... 14.
Lumbar   .do.   ... 6.
Sacral   do.   ... 3.
Caudal   do.   ... 18 (the rest are wanting).

1091. Very young skeleton. From the Zoological Society's Gardens.
                                Presented by Prof. Clark.

Cebus (species uncertain).

1092. Adult skull. The canines are short, but strong; the parietal bones join the malar bones. The intermaxillary facial suture is obliterated. [Bell collection.] Presented by Prof. Clark.

1093. Younger skull, to judge from the open state of the cranial sutures. The suture between the intermaxillary and maxillary bone is obliterated. [Bell collection.] Presented by Prof. Clark.

1094. Younger skull, with the same characters. [Bell collection.]
Presented by Prof. Clark.

1095. Skull of an animal with the same characters as the last. From the Zoological Society's Gardens. Presented by Prof. Clark.

1096. Young skull. Presented by Prof. Clark.

1097. Adult skull. The dentition is complete. [Bell collection.]
Presented by Prof. Clark.

Genus *Ateles*. GEOFFROY ST HILAIRE. **Spider Monkeys.**

Dental formula the same as the last.

Ateles (hybridus?)

Cervical vertebræ... 7.
Dorsal   do.   ... 14.
Lumbar   do.   ... 4.
Sacral   do.   ... 3.
Caudal   do.   ... 29 (a few are wanting).

1098. Young skeleton. Macartney collection.

Phalanx III. HEOPITHECI.

Genus *Cynocephalus*. CUVIER.

Dental formula: $i.\ \frac{2-2}{2-2},\ c.\ \frac{1-1}{1-1},\ p.\ \frac{2-2}{2-2},\ m.\ \frac{3-3}{3-3}=32.$

Cynocephalus porcarius. DESM. **The Chacma.**

Cervical vertebræ... 7.
Dorsal   do.   ... 13.
Lumbar   do.   ... 6.
Sacral   do.   ... 3.
Caudal   do.   ... 18 (the rest are wanting).

1099. Skeleton. From the Zoological Society's Gardens.

Presented by Prof. Clark.

1100. Head. [Bell collection.] Presented by Prof. Clark.

Cynocephalus Maimon. Cuv.

1101. Young head, taken from a stuffed specimen; as were the following bones.

1102. Right scapula.

1103. Left scapula.

1104. Right ulna.

1105. Left ulna.

1106. Right femur.

1107. Right humerus.

1108. Left humerus.

1109. Left radius. Harwood collection.

1110. Cast of a very old head. [Bell collection.]

Presented by Prof. Clark.

Cynocephalus (species uncertain).

1111. Young head. [Bell collection.] Presented by Prof. Clark.

Genus *Cercocebus.* GEOFFROY ST HILAIRE.

Dental formula the same as the last.

Cercocebus fuliginosus. GEOFFR. **The Mangabey.**

Cervical vertebræ...	7.
Dorsal do.	... 12.
Lumbar do.	... 7.
Sacral do.	... 3.
Caudal do.	... 22.

1112. Skeleton. From the Zoological Society's Gardens.

Presented by Prof. Clark.

1113. Head. [Bell collection.] Presented by Prof. Clark.

1114. Head, wanting the lower jaw. [Bell collection.]

Presented by Prof. Clark.

Cercocebus Æthiops.   **The White-crowned Mangabey.**

1115.   Old head.   [Bell collection.]   Presented by Prof. Clark.

Cercocebus (species uncertain).

1116.   Cranium : opened by a horizontal section, to show the interior of the cavity.   [Bell collection.]   Presented by Prof. Clark.

Genus *Macacus.* CUVIER.

Dental formula the same as the last.

Macacus cynomolgus.   DESM.   **The Macaque.**

> Cervical vertebræ ... 7.
> Dorsal   do.   ... 13.
> Lumbar   do.   ... 7.
> Sacral   do.   ... 2.
> Caudal (wanting).

1117.   Skeleton.   Presented by the Master and Fellows of Trinity College.

1118.   Very old head.
1119.   Old head.   } [Bell collection.]   Presented by Prof. Clark.
1120.   Adult head.

1121.   Adult head, probably of a female.   Presented by Prof. Clark.

1122.   Young head, from the Zoological Society's Gardens.
                                        Presented by Prof. Clark.

1123.   Young head, with the cranium laid open.
                                        Presented by Prof. Clark.

Macacus Nemestrinus.   DESM.   **The Pig-tailed Monkey.**

1124.   Head.   [Bell collection.]   Presented by Prof. Clark.

Macacus sinicus. DESM. **The Bonnet Macaque.**

Cervical vertebræ ... 7.
Dorsal   do.   ... 12.
Lumbar   do.   ... 7.
Sacral   do.   ... 3.
Caudal   do.   ... 18.

1125. Skeleton. From the Zoological Society's Gardens.
Presented by Prof. Clark.

1126. Skeleton. Brookes collection.

Macacus (Nemestrinus)?

1127. Head. [Bell collection.] Presented by Prof. Clark.

Macacus (species uncertain).

1128. Young skeleton, female (?), imperfect; from the Zoological Society's Gardens. Presented by Prof. Clark.

1129. Adult head. [Bell Collection.] Presented by Prof. Clark.

Genus *Cercopithecus.* ERXL.

Dental formula the same as the last.

Cercopithecus cynosurus. **The Malbrouck Monkey.**

Cervical vertebræ ... 7.
Dorsal   do.   ... 13.
Lumbar   do.   ... 7.
Sacral   do.   ... 3.
Caudal   do   ... 19. (the rest are wanting).

1130. Skeleton. From the Zoological Society's Gardens.
Presented by Prof. Clark.

Cercopithecus sabæus. F. CUVIER. **The Green Monkey.**

1131. Head dried, to show the buccal pouches.
Macartney collection.

1132. Adult head. ⎫
1133. Younger head. ⎬ [Bell collection.] Presented by Prof. Clark.

1134. Young head. From the Zoological Society's Gardens.

Presented by Prof. Clark.

Genus *Semnopithecus.* F. CUVIER.

Dental formula the same as the last.

Semnopithecus entellus. CUVIER.

Cervical vertebræ ... 7.
Dorsal   do.   ... 12.
Lumbar   do.   ... 7.
Sacral   do.   ... 3.
Caudal   do.   ... 23 (a few are wanting).

1135. Skeleton. [Bell collection.] Presented by Prof. Clark.

Genus *Simia.* ILLIGER.

Dental formula the same as the last.

Simia Satyrus. L. **The Orang Outan.**

1136. Head.
1137. Os Pubis, Illium, Ischium, of the right side.
1138. Os Pubis, Illium, Ischium, of the left side.
1139. Right scapula. The coracoid process is still distinct.
1140. Left scapula.
1141. Right femur. There is no pit for the attachment of the ligamentum teres.
1142. Left femur.
1143. Right humerus.
1144. Left humerus.
1145. Right tibia.
1146. Left tibia.
1147. Right ulna.
1148. Left ulna.
1149. Right radius.

1150. Left radius.

1151. Left fibula.

1152. Right clavicle.

1153. Left clavicle.

1154. Three sacral vertebræ.

1155. Four lumbar vertebræ.

1156. Ten dorsal vertebræ.

1157. Ribs of the left side, twelve in number.

1158. Ribs of the right side, wanting the second, the ninth, and the twelfth.

1159. The Os calcis and astragalus, right side.

1160. Sixteen separate bones of an extremity, difficult to determine on account of the absence of the epiphyses of many of them.

Presented by Prof. Clark.

The bones of this animal, No. 1136—1160 were sent in a box to Prof. Clark by an unknown contributor.

Genus *Troglodytes.* GEOFFROY ST HILAIRE.

Troglodytes Gorilla. SAVAGE. **The Gorilla.**

1161. Cast of the skull of the old male Gorilla, brought by M. du Chaillu from Equatorial Africa, and now in the British Museum.

Presented by J. W. Clark, M.A.

Order XII. BIMANA.

Family XLVI ERECTA.

Genus *Homo.* LINNÆUS.

Dental formula the same as the last.

Homo sapiens. L. **Man.**

Cervical vertebræ ...		7
Dorsal	do.	... 12
Lumbar	do.	... 5
Sacral	do.	... 5
Caudal	do.	... 3

MELANIAN (DARK BROWN OR BLACK) VARIETY.

1162.   Skeleton of a male Bosjesman.

Purchased by the University of M. Dumoutier.

1163.   Skeleton of a female Bosjesman.

Presented by W. W. Fisher, M.D.

The three following skulls of Negroes were presented by George Budd, M.D. Caius College. He describes them as follows: "They were taken from subjects who died, while under my care, in the Seamen's Hospital, Dreadnought. My case book furnishes me with the following particulars respecting them."

1164.   "(A) *Robin*, age 21; height, 5 feet, 8 inches; hair very short, woolly, and frizzled; scarcely any beard or whisker; forehead *not* reclining; nose less broad and lips less thick than is usual in negroes; front teeth filed; figure generally well formed, muscular; thighs very muscular, calves proportionally much less so, feet large and flat.

"He was a native of Rio Pongo, which is, I believe, between Sierra Leone and the Gambia, and which he had quitted for the first time in making this voyage to England. He spoke English very imperfectly, and died of cholera a few days after his admission into the Dreadnought, so that I had no opportunity of making any observations on his character. His physiognomy, with the exception of his short frizzled hair and very black skin, did not differ much from that of European races. You will see that the skull does not exhibit in a marked degree the peculiarities of the negro formation. I have met with one or two other natives of the same part of Africa, whose physiognomies did not materially differ from that of Robin. Dr Pritchard mentions, on the authority of Adanson and other travellers, the resemblance in features which natives of this part of Africa bear to Europeans, and their general superiority over negroes of the Guinea coast. In the dissection of Robin, I noticed a peculiarity in the fat, which was of a deep orange color; this color must have been natural, as there was no jaundice or yellowness of the conjunctiva. I have since met with another instance, also in an African, in which the fat was of this color."

1165.   "(B) *Tomes Martins*, age 37; height 5 feet, 8 inches; of an extremely powerful figure; face very characteristic of the negro, nose broad, lips thick, hair short and woolly, eyes large and animated, teeth filed. He died of phthisis, and was for a long time under my observation. He was one of the crew of a Portuguese slaver, which was captured by the Boneta on the Western coast of Africa, where he was employed to enslave his countrymen. He was very intelligent, spoke Portuguese fluently, and had learnt a little English, and altogether was one of the finest specimens of the negro I have ever seen. His expression was singularly rich and animated, in a degree which is never equalled in any other negro race. He was a native of Congo. I learnt from Martins that the custom of filing the teeth, which is common to many African tribes, is continued from a superstition that they are protected by it from slavery."

N.B. The great ala of the sphenoid does not meet the parietal bone.

1166. " (C) This skull was preserved in consequence of its presenting the negro characteristics of feature in a striking degree. These are still indicated by the form of the skull, which contrasts strongly with (A) and even with (B). I have mislaid the particulars of his history, but if I recollect aright, he was a native of Guinea."

1167. Models representing the physiognomy of the Hottentot, Kaffir, and Bosjesman races.

Presented in 1856 by Richard Okes, D.D. Provost of King's College, Cambridge.

1168. Head of a native Australian chief. The teeth appear to have been filed. The sutures are partially obliterated. Brought from New South Wales by Captain Saunders, who was employed by the Government to take out convicts.

Presented by George Budd, M.D. Caius College.

1169. A head of one of the Aborigines of New South Wales. It was brought by Dr Stanger, who subsequently distinguished himself in the Niger expedition, from a burying ground of the aborigines near Newcastle, in New South Wales. In obtaining this, and three other skulls, he was exposed to great peril from the natives. Presented by George Budd, M.D. Caius College.

1170. Part of the head of a native of Van Diemen's Land.

Presented by Charles Harrison, Esq. then of the Treasury.

AMERICAN RACES.

1171. Cast of the head of a Carib. Purchased of De Ville.

Presented by Prof. Clark.

1172. Head of a Greenlander: remarkable for a great projection of the malar bones, with large temporal processes. Macartney collection.

ASIATIC RACES.

1173. Head of a Chinese pirate, decapitated at Canton.

Presented by Mr Vachell.

1174. Skull reported to be of a Hindoo, probably a female. There is a beautiful set of teeth in the upper jaw. The inferior margin of the malar bone on the left side formed of a distinct piece, united to the body of the bone by suture. The ossa triquetra are very numerous. Bought of I. Deck, Chemist, Cambridge.

Presented by Prof. Clark.

EGYPTIAN RACE.

1175. A Mummy, presented to the University by the Hon. George Townshend. It was for very many years in the University Library. A description of it is given by Conyers Middleton in his "Antiquitates Middletonianæ," p. 251 sq. The incisor teeth are remarkable for their crowns formed like molars. *Vide* also Blumenbach, "Decas Collectionis suæ Craniorum", p. 14. Some of these are still visible, but the lower jaw has been removed, probably since the publication of Blumenbach's work.

1176. Head of a gilded Mummy, from the Necropolis of Thebes, Upper Egypt.

Presented by John Anthony, M.D. Caius College.

PHŒNICIAN(?) RACE.

1177. A skull of a Guanche, or aboriginal of Teneriffe.

1178. A cranium.

1179. A cranium, more perfect.

The following bones were found with the crania.

1180. Fragment of lower jaw.

1181. Right side of lower jaw.

1182. The right humerus.

1183. The fourth dorsal vertebra.

1184. The eleventh dorsal vertebra.

1185. The second rib, left side.

1186. The third rib, right side.

1187. The right femur.

1188. The right tibia.

1189. A fragment of the right radius.

1190. Lower extremity of the right ulna.

The above crania and bones were presented by Derwent Henry Smith, Esq. Port Oratava, Teneriffe.

EUROPEAN RACES.

1191. Skull found at Comberton, Cambridgeshire, near a Roman Villa, and under some Roman Pottery.

Presented by G. M. Humphrey, M.D. Downing College.

1192. A human skull found in a barrow on Bincombe down, near Weymouth, Dorset. The barrows are numerous, and supposed to be of early date. In many or most of them no skeletons are found, only remnants of cremation. Of eleven opened by the Rev. J. J. Smith, entire skeletons were found only in two.

Presented by the Rev. J. J. Smith, Caius College.

Roman.

1193. The head of a body found in a tumulus on Eastlow Hill, Rougham, Suffolk, July 4, 1844. Within the skull is a coin found in the mouth. Presented by Prof. Henslow.

[See Prof. Henslow's pamphlet, published in the *Bury Post;* and dated July 12, 1844].

Romano-British.

1194.　Three skulls found in a Romano-British burial-ground at Felixstow, Suffolk.

> N.B.　Roman coins and vases, &c. are found on the same site.
>
> Presented by Prof. Henslow.

1195.　Ancient head, found at Edix Hill, Barrington, Cambridgeshire.　Presented by W. H. Drosier, M.D. Caius College.

1196.　A well-marked adult femur, of the right side.

1197.　A well-marked adult femur of the left side.

1198.　Right tibia.

1199.　Left femur of another individual.

1200.　Left femur of a smaller individual.

1201.　Skull.　All the sutures, except the squamous, nearly obliterated.

1202.　Right half of a lower jaw; teeth remarkably perfect.

1203.　Fragments of the bones of Ruminants, found with the above.

These bones (Nos. 1196—1203), were found at the above locality.　"The bones lie about 18 in. deep in the clay.　The field is covered with them.　Most of the skulls bear marks of violence. The bosses of shields, spear-heads, beads and clasps, are found with them.　Some of these have been presented to the Camb. Antiq. Society, at whose Museum they may be seen."

> Presented by Capt. Bendyshe, of Barrington Hall.

German.

1204.　Skull of a soldier, reputed to have been hanged for the murder of his wife.

> Purchased of Mr Deck, and presented by Prof. Clark.

Swede.

1205.　The cranium of an adult male, in which the frontal sinusses are remarkably prominent.　Harwood collection.

French.

1206.　Skeleton of a Frenchman.　The bones are placed at distances to show the surfaces by which the component parts articulate.

> Bought by the University of M. Dumoutier, Paris.

1207. Skeleton of Madame Barré, an old subject, presenting cervical ribs. Macartney collection.

1208. Cast of the face of Madame Barré. Macartney collection.

The origin of the following, though of European race, is not known.

1209. Skeleton of an adult male.
Presented by the Master and Fellows of Trinity College.

1210. Skeleton of a male. Harwood collection.

1211. Skeleton of a male.
Presented by S. Stanley, Esq. Surgeon, Cambridge.

1212. The skeleton of an adult female. Macartney collection.

1213. Bones of a male skeleton complete. The frontal bone is divided into two parts. Presented by Prof. Clark.

1214. The bones of Abraham Green, shot by Mr Perry, of Strethall Hall, Essex, while burglariously entering his house at night.
Presented by Mr Perry.

1215. Skeleton of a young female, in the attitude of the Venus de Medici. Brookes collection.

1216. Skeleton of a young subject, articulated by the natural ligaments. Macartney collection.

1217. The skeleton of an infant at the time of birth.
Macartney collection.

1218. The head and trunk of a male. Macartney collection.

1219. Spinal column and pelvis of a male. Harwood collection.

1220. The trunk of a female; the ribs compressed by wearing stays.
Macartney collection.

1221. A section of the head and trunk in the mid plane.
Presented by G. M. Humphrey, M.D.

1222. Two human feet: the one articulated with the bones in contact: the other with the bones at distances, to show the articulating surfaces. Purchased by the University.

1223. Two human hands: the one articulated with the bones in contact: the other with the bones at distances, to show the articulating surfaces. Purchased by the University.

1224.   Pelvis of a male.   Harwood collection.

1225.   Pelvis of a male.   Brookes collection.

1226.   Pelvis of a female. ⎫
1227.   Pelvis of a female. ⎭ Harwood collection.

1228.   Head of Dr O'Connor, physician of Dublin, who bequeathed his body to Dr Macartney for dissection.*   Macartney collection.

1229.   Mask of the face of J. J. O'Connor, M.D.
                                              Macartney collection.

1230.   Head of Glorvina, Lady Morgan's "Wild Irish Girl."
                                              Macartney collection.

1231.   Plaster cast of the above.   Macartney collection.

1232.   Head of a subject from the hulks.   Presented by Prof. Clark.

1233.   Head of a subject from the hulks.   Presented by Prof. Clark.

1234.   Skull. ⎫
1235.   Skull. ⎬ Presented by the Master and Fellows of Trinity College.
1236.   Skull. ⎭

1237.   Head with the sutures obliterated.   Macartney collection.

1238.   Head of Matthew Moore, who died aged 104 years.   The sutures are persistent.   Macartney collection.

1239.   Head of Abraham Green.

1240.   Mask of the face of the above.

1241.   A cranium with ossa triquetra on each side between the sphenoid and parietal bones.   Macartney collection.

1242.   A head with longitudinal axis of great length.
                                              Macartney collection.

* Dr O'Connor's will is as follows:

"The last will and testament of J. J. O'Connor.

"In the name of God Amen. I bequeath my soul to God, my body to Dr Macartney of Trinity College for inspection, and any other use he may wish to put it to, requesting that he will serve notice on the Surgeon Genl. Dr Cheyne, Sir Arthur Clark and Dr Stokes, Junr. to attend at the inspection.

10th *June*, 1827.

                    Witness my hand and seal,
                              Js. J. O'Connor.
                    Witness John Finlay.
                              Henry Day.

1243. Head with a distinct piece on the inferior margin of the malar bone on the left side. Harwood collection.

1244. A well-shaped head, with the sutures in process of obliteration, except the squamous, and sphenoido-frontal. Macartney collection.

1245. A head with narrow forehead, and large temporal ridges and fossæ. Macartney collection.

1246. A head in which the squamous bone advances to meet the frontal bone between the sphenoid and parietal on the right side only. Harwood collection.

1247. A head in which the sutures are nearly obliterated with the exception of the squamous.

1248. Head in which the sphenoid and parietal bones meet by a very small edge. Curious ossa triquetra in the course of the lambdoidal suture. Macartney collection.

1249. Head with the sutures, except the squamous, in course of obliteration. Macartney collection.

1250. An old head with a remarkable projection of the nasal bones and nasal process of the superior maxillaries: all the sutures, except the squamous, are in course of obliteration.

Macartney collection.

1251. Head with divided frontal bone. There is an additional piece on the inferior margin of the malar bone on the right side. Numerous ossa triquetra in the course of the lambdoidal suture. Indication of the suture between the intermaxillary bones and the maxillary in the palate. Macartney collection.

1252. Head with the styloid processes of the temporal bone very large and naturally attached.

1253. Head with the alveolar processes filed down in order to show the natural position of the teeth in both jaws.

Macartney collection.

1254. A human head disarticulated: with the bones at distances to show the edges and surfaces by which they meet. The osseous organ of hearing is dissected on each side.

Bought by the University.

15

1255. Transverse section of a fragment of a churchyard skull in which the anterior, middle, and posterior clinoid processes are united by osseous matter; thus forming a foramen for the passage of the ophthalmic artery.

Presented by Rev. E. G. Jarvis, Trin. Coll.

1256. Tranverse section of a cranium, showing irregularities of form in the foramen magnum, etc.   Macartney collection.

1257. Transverse section of a cranium.   The middle fossa of the left side is unsymmetrically small.   The general form of the skull is globular.   Macartney collection.

1258. Transverse section of a cranium, showing the connection of the superior ethmoidal cells and frontal sinusses.

Macartney collection.

1259. Transverse section of a head, showing the connection between the ethmoidal and frontal sinusses.   Macartney collection.

1260. A head with the surface marked out in Phrenological regions according to the system of Spurzheim.

Presented by Mr Deck, of Cambridge.

1261. The head of a young subject, in which there are many peculiarities, particularly a division of the occipital bone by a suture which passes transversely from the posterior angle of the temporal bone to the corresponding angle in the other.

Harwood collection.

1262. The head of a young child, with the bones connected at distances, to show their relations to each other.

Presented by Prof. Clark.

1263. Young head with projecting occiput and large ossa triquetra in the course of the lambdoidal suture.   Macartney collection.

1264. Mask of the face of Sir Isaac Newton.

Presented by the Syndics of the University Library.

1265. Mask of the face and neck of the Right Honourable William Pitt.   Presented by the Syndics of the University Library.

1266. Mask of the face of the Right Honourable James Fox.

Presented by the Syndics of the University Library.

1267. Mask of the face and neck of the Right Honourable Spencer Perceval.   Presented by the Syndics of the University Library.

1268.   Mask of the face of Charles XII.   See note at the end of the volume.   Presented by the Syndics of the University Library.

1269.   Mask of the face of Benjamin Franklin; purchased of De Ville.
                                                Presented by Prof. Clark.

1270.   Bust of the late Charles Matthews, Esq. Comedian.
                                         Presented by Mr S. Pryor, Jun. Cambridge.

1271.   The cast of a well-shaped head, from De Ville's collection.
                                                Presented by Prof. Clark.

1272.   A similar head, from the same.   Presented by Prof. Clark.

1273.   Plaster cast of a skull (No. 1231).
                          Presented by the Master and Fellows of Trinity College.

1274.   Cast of a bust, exhibiting the muscles of the face and neck.
                                                Presented by Prof. Clark.

1275.   Three Phrenological busts.
                                    Presented by Mr Deck, Chemist, Cambridge.

1276.   Cast of the head of a female who was in the habit of sticking pins into her person.   Macartney collection.

1277.   Bust of John Thurtell, who was executed at Hertford, on Friday, the 9th of January, 1824, for the murder of Mr W. Weare. Purchased of De Ville.   Presented by Prof. Clark.

1278.   Bust of Mary Mc Kenis, the Scotch murdress.   Purchased of De Ville.   Presented by Prof. Clark.

1279.   Bust of Williams, who burked the Italian boy.   Purchased of De Ville.   Presented by Prof. Clark.

1280.   Bust of Joshua Slade, who murdered Mr Waterhouse at Stukely, Hunts.   Purchased of De Ville.
                                                Presented by Prof. Clark.

1281.   Bust of J. B. Rush, who committed the murders at Stanfield Hall, Wymondham, Norfolk, in November 1848.
                                                Presented by Prof. Clark.

1282.   Cast from an antique bust of the Infant Bacchus.
                                                Presented by Prof. Clark.

1283.   Cast of a colossal arm removed from a statue found in the cave of Elephanta in India.   Macartney collection.

1284.   Cast of the arm of Mons. Huguenin, bent, with the muscles in action.   Macartney collection.

1285. Cast of the arm of an athletic man, in a state of extension.
           Macartney collection.

1286. Cast of the leg of Madame Vestris. Macartney collection.

1287. Cast of the foot of a female who never wore shoes.
           Macartney collection.

1288. A model in plaster of the left side of the body of a female, to show the form in its natural proportions. Macartney collection.

1289. A model in plaster of the same female, to show the disfigurements produced by stays, garters, and shoes. Macartney collection.

NOTE ON THE MASK OF THE FACE OF CHARLES XII.

This cast confirms so remarkably the story of the assassination of Charles, that it is worth while comparing the testimonies of the various authors who have related his death.

Voltaire, in his *Histoire de Charles XII.* (Works by Beuchot, Vol. xxiv. p. 351 sq.) says, in the first place, that on the 11th December, 1718, the King, who was besieging Frederickshall, went down to the trenches, "vers les neuf heures du soir." After noticing the story of conversations held by him with Mégret, an engineer, which he declares to be false, he resumes with "Voici ce que je sais de véritable sur cet événement. Le roi était exposé presque à demi corps à une batterie de canon pointée vis-à-vis l'angle où il était : il n'y avait alors auprès de sa personne que deux Français; l'un était M. Siquier, son aide-de-camp, homme de tête et d'exécution, qui s'était mis à son service en Turquie, et qui était particulièrement attaché au prince de Hesse ; l'autre était cet ingénieur. Le canon tirait sur eux à cartouches; mais le roi, qui se découvrait davantage, était le plus exposé. A quelques pas derrière était le Comte Schwerin, qui commandait la tranchée. Le comte Posse, capitaine aux gardes, et un aide-de-camp nommé Kaulbar, recevaient des ordres de lui. Siquier et Mégret virent dans ce moment le roi de Suède qui tombait sur le parapet en poussant un grand soupir; ils s'approchèrent; il était déjà mort. Une balle pesant une demi-livre l'avait atteint à la tempe droite, et avait fait *un trou dans lequel on pouvait enfoncer trois doigts; sa tête etait renversée sur le parapet, l'œil gauche était enfoncé, et le droit entièrement hors de son orbite.** L'instant de sa blessure avait été celui de sa mort; cependant il avait eu la force, en expirant d'une manière si subite, de mettre, par un mouvement

---

* The italics are mine.

naturel, la main sur la garde de son épée, et était encore dans cet attitude. A ce spectacle, Mégret, homme singulier et indifférent, ne dit autre chose, sinon: 'Voilà la pièce finie, allons souper.' Siquier court sur-le-champ avertir le comte Schwerin. Ils résolurent ensemble de dérober la connaissance de cette mort aux soldats, jusqu'à ce que le prince de Hesse en pût être informé."

To which narrative Beuchot appends a note: "Le procès-verbal de l'autopsie cadaverique, faite en 1746, établit que le coup qui avait traversé les deux tempes n'y avait laissé qu' une blessure longue de sept lignes, et large de deux. Une *balle d'une demi-livre* eût laissé bien d'autres traces." We may add to this that the cast proves the hole to have been that made by an ordinary pistol-bullet, measuring $\frac{7}{10}$ths of an inch in length, by $\frac{4}{10}$ths in breadth, 1 inch above the eyebrow, and 2 inches from the middle of the forehead, and that neither the eyes nor any other part of the face bears the slightest marks of violence.

A little further on in his history (p. 357) Voltaire alludes to the story of the assassination, and says that a report got abroad in Germany that Siquier had killed the King. "Ce brave officier fut long-temps désespéré de cette calomnie: un jour, en m'en parlant, il me dit ces propres paroles: 'J'aurais pu tuer le roi de Suède; mais tel était mon respect pour ce héros, que si je l'avais voulu, je n'aurais pas osé.'"

He then proceeds to defend Siquier in a passage added in 1748—his history having been first published in 1731. "Je sais bien que Siquier lui-même avait donné lieu à cette fatale accusation qu'une partie de la Suède croit encore; il m'avoua lui-même qu'à Stockholm, dans une fièvre chaude, il s'était écrié qu'il avait tué le roi de Suède; que même il avait dans son accès ouvert la fenêtre, et demandé publiquement pardon de ce parricide. Lorsque dans sa guérison il eut appris ce qu'il avait dit dans sa maladie, il fut sur le point de mourir de douleur. Je n'ai point voulu révéler cette anecdote pendant sa vie. Je le vis quelque temps avant sa mort, et je peux assurer que loin d'avoir tué Charles XII, il se serait fait tuer pour lui mille fois. S'il avait été coupable d'un tel crime, ce ne pouvait être que pour servir quelque puissance qui l'en aurait sans doute bien récompensé; il est mort très pauvre en France, et même il y a eu besoin du secours de ses amis. Si ces raisons ne suffisent pas, que l'on considère que la balle qui frappa Charles XII ne pouvait entrer dans un pistolet, et que Siquier n'aurait pu faire ce coup détestable qu'avec un pistolet caché sous son habit." To which is appended a note by the Kehl editors, MM. Condorcet and Decroix: "Beaucoup de gens prétendent encore que Charles XII. fut la victime de la haine qu'il avait inspirée à ses sujets. Cette opinion n'est.

pas même destituée de vraisemblance.　M. de Voltaire ne l'ignorait pas ; mais comme il ne pouvait vérifier les petites circonstances sur lesquelles cette opinion s'appuie, il a préféré la passer sous silence.　On garde à Stockholm le chapeau de Charles XII ; et la petitesse du trou dont il est percé est une des raisons de ceux qui veulent croire qu'il périt par un assassinat."

The appearance of the cast seems to settle the question in favour of the story that Charles was murdered : a story which is rendered all the more probable by the fact of there being then two parties in the kingdom respecting the succession : the one, favoured by himself, his minister Görtz, and the Court of Russia, for the young Duke of Holstein, son of his elder sister Sophia : the other for his younger sister Ulrica Eleonora, and her husband Prince Frederick of Hesse Cassel.　Her chance of success, her rival being so powerfully seconded, depended on the death of Charles, while the Duke of Holstein was still a boy, and unable to act with decision in a sudden emergency.　In fact, she was elected Queen with little opposition on the news of the King's death reaching Stockholm.

Geyer, himself a Swede, describes the death of Charles thus (French translation of his *History of Sweden*, p. 500): "Ce fut durant les travaux de ce siége que Charles fut assassiné par les siens d'un coup de pistolet, le 11 decembre, à dix heures du soir."

ENGLISH INDEX.

Cock, common, 34.
Cod, 5.
Coffer-fish, 4.
Condor, 45.
Coot, bald, 26.
Cormorant, green, 24.
Corncrake, 27.
Copyu, 77.
Crane, crested, 30.
Crocodile, common, or of the Nile, 12.
— narrow-beaked, or of the Ganges, 13.
Crow, African, 39.
— Carrion, 39.
— hooded, or Royston, 39.
Cuckoo, 37.

D.

Dab-chick, 19.
Deer, barking, 65.
— Fallow, 65.
— gigantic Irish, 66.
— red, 64.
— Rein, 64.
— Roe, 66.
Diver, black-throated, 18.
— great Northern, 19.
— red-throated, 18.
Dodo, 36.
Dog, 88.
Dolphin, 51.
— Cape, 51.
Dory, 7.
Dotterel, 31.
Dragon, 12.
Dragonet, gemmeous, 6.
Duck, Eider, 20.
— golden-eyed, 20.
— Scaup, 20.
— Shiel, 21.
— wild, 20.
Dugong, Australian, 52.

E.

Eagle, golden, 44.
— white-tailed, 44.
Echidna, short-spined, 46.
Eland, 68.
Elephant, African, 54.
— Indian, 53.
— Sea, 80.
Elk, 64.
— Irish, 66.
Emeu, 32.

F.

Falcon, Peregrine, 43.
Flamingo, 30.
Flying-fish, 5.
Foumart, 90.
Fowl, Dorking, 35.
— Guinea, 34.
— Poland, 35.
Fox, 87.
Frigate-bird, 23.
Frog, common, 9.
— edible, 8.
— green, 8.
— tree, 9.
Frog-fish, 6.
Fulmar, northern, 25.

G.

Gannet, 23.
Gar-fish, 5.
Gavial, 13.
Giraffe, 67.
Goat, 69.
Godwit, bar-tailed, 28.
— black-tailed, 28.
Goosander, buff-breasted, 19.
— red-breasted, 19.
Goose, Brent, 22.
— Egyptian, 22.

Monitor, large, 11.
Monkey, Capuchin, 99.
— green, 103.
— Malbrouck, 103.
— pig-tailed, 102.
— Spider, 100.
Morse, 80.
Mouse, common, 77.
— meadow, 78.

N.

Natter-jack, 9.
Noctule, 97.
Nylghau, 68.

O.

Ocelot, 84.
Opossum, Virginian, 48.
Orang-Outan, 104.
Osprey, 45.
Ostrich, African, 32.
— New-Holland, 32.
Otter, common, 90.
Owl, Barndoor, 43.
Ox, 70.
Oyster-catcher, 30.

P.

Paca, brown, 75.
Parakeet, rose-billed, 36.
Partridge, common, 35.
— red-legged, 35.
Peacock, 34.
Pelican, 23.
Penguin, 17.
Perch, 7.
— sea, 8.
Petrel, storm, 26.
Phalarope, grey, 28.
Pheasant, 34.
Pig, common, 60.

Pigeon, common, 36.
— wood, 35.
Pike, 5.
— sea, 5.
Pipe-fish, 3.
Plaice, 6.
Plover, golden, 31.
— gray, 31.
— Norfolk, 31.
Pochard, common, 20.
Polecat, 90.
Porcupine, Asiatic, 76.
— common or crested, 76.
— of Java, 76.
Porpoise, 50.
Potoroo, 47.
Puffin, 17.
Puma, red, 85.

Q.

Quail, 35.

R.

Rabbit, 75.
Racoon, 95.
Rasse, 87.
Rat, brown or Norway, 77.
Ratel, 93.
Raven, 38.
Ray, Shagreen, 1.
Razor-bill, 17.
Redshank, 28.
Reindeer, 64.
Rhinoceros, Indian, 55.
Robin, 42.
Rook, 39.

S.

Sagittary, Cape, 45.
Sanderling, 28.
Sandpiper, little, 28.
— purple, 28.

LATIN INDEX.

The Generic names are in ordinary type, the Specific in italics.

C.

Delphinus *Tursio*, 50.
Didelphis *virginiana*, 48.
Didus *ineptus*, 36.
Dinornis *casuarinus*, 33.
— *didiformis*, 33.
— *giganteus*, 33.
Diodon *hystrix*, 4.
Diomedea *exulans*, 26.
— *fuliginosa*, 26.
— *melanophrys*, 26.
Diprotodon, 48.
Draco *volans*, 12.
Dromaius *novæ Hollandiæ*, 32.

E.

Echeneis *remora*, 6.
Echidna *arietans*, 10.
— *setosa*, 46.
Edolius *remifer*, 42.
Elephas *africanus*, 54.
— *indicus*, 53.
Emyda *punctata*, 14.
Emys *picta*, 14.
— *rugosa*, 14.
— *tecta*, 14.
Emysaura *serpentina*, 14.
Eques, 7.
Equus *asinus*, 59.
— *caballus*, 57.
Erinaceus *europæus*, 97.
Esox *lucius*, 5.
Exocœtus, 5.

F.

Falco *Bacha*, 44.
— *Nisus*, 43.
— *peregrinus*, 43.
— *Pondicereanus*, 43.
— *tinnunculus*, 43.
Felis *catus*, 85.
— *concolor*, 85.
— *Javanensis*, 85.
— *Leo*, 83.

Felis *Leopardus*, 84.
— *pardalis*, 84.
— *Tigris*, 84.
Fringilla *canarina*, 41.
— *chloris*, 40.
— *domestica*, 40.
— *cœlebs*, 40.
Fulica *atra*, 26.
Fuligula *marila*, 20.

G.

Gadus *morrhua*, 5.
Galeopithecus *varius*, 98.
Galictis *vittata*, 91.
Gallinula *chloropus*, 27.
— *phœnicurus*, 27.
Gallus *cristatus*, 35.
— *furcatus*, 35.
— *gallorum*, 34.
— *pentadactylus*, 35.
Glyptodon *clavipes*, 73.
Grus *pavonina*, 30.
Gypogeranus *serpentarius*, 45.

H.

Hæmatopus *ostralegus*, 30.
Haliaëtus *albicilla*, 44.
Halichærus *gryphus*, 81.
Halicore *australis*, 52.
Helarctos *Tibetanus*, 94.
Herpestes *Ichneumon*, 87.
Hippocampus *brevirostris*, 4.
Hippoglossus *vulgaris*, 5.
Hippopotamus *amphibius*, 61.
Hirundo *riparia*, 42.
— *rustica*, 42.
— *urbica*, 42.
Homo *sapiens*, 105.
Hyæna *spelæa*, 86.
— *striata*, 85.
Hydraspis, 14.
Hydrochœrus *capybara*, 75.

Strix *Ceylonensis*, 43.
— *flammea*, 43.
Strongyloceros *spelæus*, 67.
Struthio *camelus*, 32.
Sturnus *vulgaris*, 40.
Sula *Bassana*, 23.
Sus *Babyrussa*, 61.
— *scrofa*, 60.
Sylvia *hortensis*, 42.
— *rubecula*, 42.
Syngnathus, 3.

T.

Tachypetes *aquilus*, 23.
Tadorna *vulpanser*, 21.
Talpa *europæa*, 96.
Tamias *striatus*, 79.
Tapirus *americanus*, 56.
Testudo *actinodes*, 15.
— *carbonaria*, 15.
— *græca*, 15.
— *indica*, 15.
— *pardalis*, 15.
Tetrodon, 4.
Thalassidroma *pelagica*, 26.
Totanus *affinis*, 28.
— *calidris*, 28.
Trichecus *Rosmarus*, 80.
Trigla *gurnardus*, 7.
Tringa *maritima*, 28.
— *minuta*, 28.
Trionyx *gangeticus*, 13.
— *labiatus*, 13.
Trochilus, 38.
Troglodytes *Gorilla*, 105.
Tropidonotus *torquatus*, 10.
Turdus *macrourus*, 41.
— *merula*, 41.

U.

Uria *Brünnichii*, 18.
— *Grylle*, 18.
— *lacrymans*, 18.
— *Troile*, 18.
Ursus *Americanus*, 93.
— *Arctos*, 93.
— *labiatus*, 94.
— *maritimus*, 94.
— *spelæus*, 95.

V.

Vanellus *cristatus*, 31.
— *tricolor*, 31.
Varanus *Bengalensis*, 11.
— *Niloticus*, 11.
Vespertilio *noctula*, 97.
— *pipistrellus*, 98.
Vipera *berus*, 9.
Viverra *Genetta*, 86.
— *malaccensis*, 87.
Vultur, 45.

X.

Xiphias *gladius*, 7.

Z.

Zeus *faber*, 7.
Ziphius *Sowerbiensis*, 49.

CAMBRIDGE: PRINTED AT THE UNIVERSITY PRESS.

Printed in the United States
By Bookmasters